Praise for *Ins*

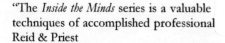

"Need-to-read inside information and ... n
line—the best source in the business." - ... h
LLP

"The *Inside the Minds* series is a valuable ... d
techniques of accomplished professional ... n
Reid & Priest

"Aspatore has tapped into a goldmine of knowledge and expertise ignored by other publishing houses." – Jack Barsky, Managing Director, Information Technology and CIO, ConEdison Solutions

"Unlike any other publisher—actual authors that are on the front lines of what is happening in industry." – Paul A. Sellers, Executive Director, National Sales, Fleet and Remarketing, Hyundai Motor America

"A snapshot of everything you need..." – Charles Koob, Co-Head of Litigation Department, Simpson Thacher & Bartlet

"Everything good books should be—honest, informative, inspiring, and incredibly well written." – Patti D. Hill, President, BlabberMouth PR

"Great information for both novices and experts." – Patrick Ennis, Partner, ARCH Venture Partners

"A rare peek behind the curtains and into the minds of the industry's best." – Brandon Baum, Partner, Cooley Godward

"Intensely personal, practical advice from seasoned deal-makers." – Mary Ann Jorgenson, Coordinator of Business Practice Area, Squire, Sanders & Dempsey

"Great practical advice and thoughtful insights." – Mark Gruhin, Partner, Schmeltzer, Aptaker & Shepard PC

"Reading about real-world strategies from real working people beats the typical business book hands down." – Andrew Ceccon, CMO, OnlineBenefits Inc.

"Books of this publisher are syntheses of actual experiences of real-life, hands-on, front-line leaders—no academic or theoretical nonsense here. Comprehensive, tightly organized, yet nonetheless motivational!" – Lac V. Tran, Senior Vice President, CIO, and Associate Dean, Rush University Medical Center

"Aspatore is unlike other publishers...books feature cutting-edge information provided by top executives working on the front lines of an industry." – Debra Reisenthel, President and CEO, Novasys Medical Inc.

www.Aspatore.com

Aspatore Books is the largest and most exclusive publisher of C-Level executives (CEO, CFO, CTO, CMO, partner) from the world's most respected companies and law firms. Aspatore annually publishes a select group of C-Level executives from the Global 1,000, top 250 law firms (partners and chairs), and other leading companies of all sizes. C-Level Business Intelligence™, as conceptualized and developed by Aspatore Books, provides professionals of all levels with proven business intelligence from industry insiders—direct and unfiltered insight from those who know it best—as opposed to third-party accounts offered by unknown authors and analysts. Aspatore Books is committed to publishing an innovative line of business and legal books, those which lay forth principles and offer insights that, when employed, can have a direct financial impact on the reader's business objectives, whatever they may be. In essence, Aspatore publishes critical tools—need-to-read as opposed to nice-to-read books—for all business professionals.

Inside the Minds

The critically acclaimed *Inside the Minds* series provides readers of all levels with proven business intelligence from C-Level executives (CEO, CFO, CTO, CMO, partner) from the world's most respected companies. Each chapter is comparable to a white paper or essay and is a future-oriented look at where an industry/profession/topic is heading and the most important issues for future success. Each author has been carefully chosen through an exhaustive selection process by the *Inside the Minds* editorial board to write a chapter for this book. *Inside the Minds* was conceived in order to give readers actual insights into the leading minds of business executives worldwide. Because so few books or other publications are actually written by executives in industry, *Inside the Minds* presents an unprecedented look at various industries and professions never before available.

Construction Law Client Strategies

*Leading Lawyers on Drafting Proper Contracts,
Understanding Legal Rights and Remedies,
and Having Success on Behalf of Clients*

ISBN 978-1-59622-640-1
Library of Congress Control Number: 2007920836

For corrections, updates, comments or any other inquiries please email editorial@aspatore.com.

First Printing, 2007
10 9 8 7 6 5 4 3 2 1

Construction Law
Client Strategies

Leading Lawyers on Drafting Proper Contracts, Understanding Legal
Rights and Remedies, and Having Success on Behalf of Clients

CONTENTS

Building Successful Client Relationships

Trent A. Gudgel

Shareholder

Hall, Estill, Hardwick, Gable, Golden & Nelson PC

Construction law has several components that overlap and fall within other defined areas of the law. It is, to a large extent, a specialized type of contract law. Construction contracts are specialized agreements that contain provisions that, like all contracts, are subject to the laws regarding contract interpretation. Construction contracts are also subject to the basic requirements of contract law and the five elements of an enforceable contract: offer, acceptance, consideration, legal capacity, and lawful purpose. Courts universally recognize that construction contracts can impose implied obligations upon the parties as well as the express obligations contained in the written documents. The implied obligations assumed by parties to a construction contract include the duty of good faith and fair dealing and the duty not to unreasonably interfere with the progress of the work.

Construction contracts come in all shapes and sizes, and they are as varied as the projects to which they relate. Several construction industry organizations have developed standard form contracts and other construction documents. These organizations include the Associated General Contractors of America, the American Institute of Architects, the Engineers Joint Contract Document Committee, the Construction Specialization Institute, and the federal government. The contract forms and construction document forms developed by these organizations overlap in many ways, with slight variations that tend to favor the party (i.e., general contractor, architect, engineer) these organizations sponsor. There are also various types of contracts used in the construction industry to deliver the project to the owner. Some of these contract types include design/build contracts; construction management contracts; engineering, procurement, and construction contracts; "cost plus" contracts; guaranteed maximum price contracts; and fixed price contracts, to name a few. Each of these contract types can be modified to be more beneficial to the owner or to the contractor.[1]

As a construction lawyer, I represent contractors, subcontractors, owners, engineers, architects, and suppliers in all aspects of construction project planning, document preparation, and dispute resolution. I provide legal

---

[1] Attached as Appendix A is a sample contract for engineering, procurement, and construction services that is pro-owner, and attached as Appendix B is a sample power plant construction contract that is pro-contractor.

advice and guidance in solving problems related to contract terms and conditions, contract requirements and procedures, statutory rights and remedies, and project dispute resolution and strategies. I also advise clients on the meaning and significance of various contract terms, clauses, and provisions, and how particular provisions can impact my client's rights and duties on the project. Lastly, I provide advice regarding strategies for resolving disputes involving project design, payment, project termination, property damage, and personal injury.

The Major Issues of Construction Law

I devote a great deal of my time to helping clients with legal issues relating to contract terms and conditions. I help clients understand the scope and potential impact of contract clauses including indemnity provisions, *force majeur* provisions, default and notice of default provisions, termination provisions, claim submission procedures, change order provisions, damage provisions (including liquidated damages, consequential damages, damage limitations, and delay damages), payment provisions, insurance requirements, and dispute resolution provisions. Many of these contract clauses are unique to construction projects, and a body of law interpreting these unique clauses has developed over the years. Many of these unique contract clauses are illustrated in the attached contract forms.[2] By understanding the various terms and conditions of their agreements, clients can better negotiate the terms of the contract, accept the contract with a better understanding of their obligations, or choose not to participate in the project because the risks imposed by the contract are too great.

In the area of project dispute resolution, I assist clients in preserving their rights through filing mechanics' and materialmen's liens, initiating litigation to enforce legal rights and recover monetary damages, and defending against claims seeking damages. The major dispute areas I help clients address involve project completion, delays, or acceleration; changes to the work; payment for work; and defective design. I help clients on both sides of these issues work through their problems and achieve a satisfactory resolution.

---

[2] See sample contract for engineering, procurement, and construction services (Appendix A), Article V, Article VIII, Article X, Article XI, and Article XVIII, and sample power plant construction contract (Appendix B), paragraphs 6A–d, 9, 10, 15, 21, and 23A–e.

Issues regarding the completion of work typically involve interference with the work, delays in completion of the work, acceleration of the work, and the costs associated with these issues. Issues relating to payment for work arise out of disputes over the scope of the project, changes and additions to the scope of the project, the quality of the work, the cost of the work, and the ability of the owner or contractor to pay for the work. Payment issues often go hand-in-hand with changes to the work, the scope of those changes, and the cost of those changes. Design issues can involve claims of defective design that have resulted in personal injury or property damage and claims that the facility does not perform as promised. Well-drafted contracts address each of these issues and provide clear procedures for resolving them.

Many construction law issues arise in the area of mechanics' and materialmen's liens. These special liens allow parties that have provided labor and materials for the improvement of real property to obtain a lien against the real property, which may be foreclosed to satisfy payment for the value of the labor and materials provided.[3] Disputes often arise regarding a party's right to enforce a mechanics' and materialmen's lien and whether a party has complied with notice requirements and other requirements of this specialized area of the law. Another component of construction law is tort law, which often involves issues of liability for design and product defects.

The Role of the Construction Lawyer: Service

In the area of construction contract law, I assist clients in drafting and understanding construction contracts and related documents that control their duties, rights, and obligations in the performance of a construction project. It is important and necessary for construction lawyers to advise clients regarding the meaning and effect of particular contract provisions on their performance of work on the project. I strive to make clients aware of the potential risks and rewards associated with particular contract provisions and how each contract clause can come into play during the course of the project. Clients may be unaware that certain terms of the

---

[3] Appendix C is an Oklahoma form of subcontractor mechanics' and materialmen's lien. Appendix D is an Oklahoma form of subcontractor pre-lien notice.

contract are subject to interpretation, and reasonable minds may differ on the specific meaning of particular provisions of the contract. I inform clients that under certain circumstances contract provisions may be considered ambiguous and, if a dispute arises, it may be necessary for a jury or another third party to decide what the parties intended when they agreed to the terms of the contract.

In the area of statutory mechanics' and materialmen's lien law, I advise clients on the rights of mechanics' and materialmen's lien claimants to impose liens upon a project in order to obtain payment for labor and materials provided. I provide guidance and counseling on the technical procedures that must be followed to obtain and enforce a mechanics' lien and the procedures that must be followed to foreclose the mechanics' lien through litigation.

In the area of design defects and negligence, I counsel clients on how to minimize risks of exposure to claims for personal injury and property damage. Owners and contractors typically minimize their risk of exposure for negligent design by including insurance, indemnification, and hold harmless provisions in their contract documents.[4] The purpose of these provisions is to shift the risk of loss for certain acts of negligence from the owner to the contractor or other parties.

Adding Direct Value for the Construction Law Client

Most commercial construction projects are a significant investment of money, time, and resources, and clients understandably have a desire to fulfill the requirements of any construction project as efficiently as possible. Construction lawyers add the most value to a construction client in the areas of contract drafting and negotiation and dispute negotiation and resolution. In the construction project documentation phase, it is important to focus on drafting documents that achieve the general and specific goals of the parties. Well-drafted project documents should allow the parties to deliver the project and achieve their business objectives while at the same time protecting the client's interest and limiting its exposure to the risks

---

[4] See sample contract for engineering, procurement, and construction services (Appendix A), Article XIV and Article XV, and sample power plant construction contract (Appendix B), paragraphs 23A–e and 22.

associated with the project. During the project planning and document drafting phase, I help clients identify and understand complex provisions of the contract documents that may create additional expense and risk exposure the client does not necessarily have to assume. I help clients develop and negotiate alternative contract terms that reduce their risk and are acceptable to all parties.

Throughout the contract drafting process, construction lawyers should strive to help clients identify the key business terms they must obtain to have a successful project and make them aware of the scope and degree of the responsibility they will assume by signing the contract documents. Often, clients are unaware of their responsibilities. For example, many times clients are unaware of the indemnification obligations they are assuming on a particular project, including indemnification of a third party's sole negligence. Also, contractors may be asked to release their lien rights, and subcontractors may be asked by the general contractor to rely solely on the owner's ability to pay as a guarantee of payment even though the subcontractor does not have a contract with the owner. By identifying and negotiating indemnification provisions, payment provisions, and other provisions of contracts before the start of a project, clients are able to avoid potentially significant costs at the end of a project.

In the litigation area, construction lawyers add value through advising clients of the legal rights and remedies they possess and the procedures they must follow to enforce their rights. I often provide advice regarding case law and other legal precedents that affect a client's ability to pursue claims or defend claims arising out of a construction project. The legal knowledge I offer includes knowledge regarding common law and statutory law relating to limitations of damages, foreclosure of mechanics' liens, and the enforceability of certain contract provisions that are unique to the construction industry. Finally, in any construction dispute it is important to provide a cost/benefit analysis and identify the strengths and weaknesses of the client's legal positions. By advising clients in this manner and identifying the risks and potential rewards, clients are better able to make decisions that will allow them to achieve a successful outcome.

Staying out of Trouble: Common Pitfalls and Problem Areas

Clients frequently get into trouble by failing to read and understand the terms of the contract they have signed. Clients often assume certain terms and provisions are included or excluded from a contract when in fact they are not, or they assume a provision has a particular meaning when in fact it means something else. Generally speaking, commercial construction contracts require any changes to the project to be in writing and approved by the parties before either party is entitled to enforce the change and before the party providing the additional work or materials is entitled to seek payment for the change. Commonly, clients may agree in principle to changes to the project and then agree to "settle up" on the costs or time associated with the change at the end of the project, after the additional work has been completed. This often results in a situation where a project owner has received additional work but does not believe the additional work is worth the amount the contractor is seeking as compensation, or the additional work has taken far longer to complete than the parties anticipated. When the parties fail to reduce the terms and provisions of the changes in the work to writing, including the specific scope of the additional work, the cost for the additional work, and time to complete the project, disagreements can arise at the conclusion of the project, which often result in litigation or other dispute resolution proceedings.

Most construction contracts contain provisions that set forth how changes to the scope of the project must be approved.[5] If these procedures are not followed, it is difficult for a party seeking additional compensation to prevail on a claim at the end of a project. A party seeking payment for additional work may be able to rely on principles of *quantum meruit*, [6] but that places an additional evidentiary burden on the parties that could be avoided through written agreement. Most construction contracts also contain provisions that require parties to give notice of changes in the

---

[5] See sample contract for engineering, procurement, and construction services (Appendix A), Article 8, and sample power plant construction contract (Appendix B), paragraphs 15A–d.

[6] *Quantum meruit* is a term used to describe an equitable doctrine based upon the concept that no one who benefits by the labor and materials of another should be unjustly enriched thereby. Under such circumstances, the law implies a promise to pay a reasonable amount for the labor and materials furnished.

condition of the work and notice of interference with the work that may result in additional expense to a contractor or subcontractor.[7] If the requirements of the contract regarding notice of potential claims are not followed, it may create a situation where a party who has been injured due to acts beyond its control is not entitled to seek compensation for its injuries.

An area of construction law that frequently results in problems for clients—particularly general contractors, subcontractors, and suppliers—is the area of mechanics' lien law. Mechanics' liens and the remedies that flow from mechanics' liens are creatures of statute. The statutory schemes for mechanics' liens are specific and require lien claimants to follow certain procedures to enforce a lien against the property. Mechanics' lien statutes prohibit a party from enforcing a mechanics' lien against the project if the statutory requirements have not been met. Frequently, clients run into trouble in this area by failing to give timely notice of a mechanics' lien claim, failing to timely file a mechanics' lien with the appropriate public authority, or failing to take other steps necessary to perfect such a lien in a timely fashion. Most mechanics' lien laws require lien claimants to provide notice of their lien claim within a certain number of days. Mechanics' lien laws also require lien claimants to file, with the county clerk of the county in which the project is situated, a written statement setting forth specific information about the mechanics' lien and describing the property on which the lien is claimed. The mechanics' lien is a powerful tool available to contractors, subcontractors, and suppliers to assist them in obtaining payment for labor and materials provided to improve real property. However, if a lien claimant does not follow the statutory procedures, the benefit of this tool can be lost.

To avoid these and similar problem areas, I always advise my clients to spend time at the beginning of the project to become familiar with the requirements of the project documents and to seek the advice of counsel regarding the terms of the contract and other documents before signing them. I also advise clients to be familiar with (and I assist them in becoming familiar with) the mechanics' lien laws in the jurisdiction where the project is located, as not all mechanics' lien laws are the same. Clients are dollars

---

[7] See sample power plant construction contract (Appendix B), paragraphs 16 and 21.

ahead if they spend time at the beginning of the project addressing these issues before a problem arises.

The Successful Construction Lawyer: Strategy and Methodology

Many construction-related issues can be avoided or minimized by aggressively addressing the issue when it first arises. For example, scope of work, changes to the work, and payment issues can be minimized through diligent contract administration and written communication with the parties involved. Issues often arise during the course of a project regarding differences between the original scope of the work and the work that is being requested. If a party believes the requested work is beyond the original scope of work, the party should communicate these differences in writing to make clear their position regarding the differences. The party should also refer to the contract to determine what procedures should be followed to preserve any claim that may arise. If a dispute later arises, it is always beneficial to have a written record of communications that occurred during the course of the project. Dispute resolution usually happens many months after the issue arises, when memories have faded and the parties' positions have hardened. Contemporaneous communications are often the most reliable source for identifying the positions of the parties at the time.

When I am asked to assist clients on construction-related issues, my strategy starts with obtaining all relevant documents, including contemporaneous communications, and identifying persons with knowledge of facts relating to the issues. After obtaining and reviewing the relevant documents, I communicate with the client representatives and friendly third parties involved to get a better understanding of the issues involved. A full understanding of the facts relating to a construction issue is crucial to providing meaningful and accurate legal advice, and it assists in resolving the issue.

Success in my role as legal counsel is predicated upon having an understanding of my client's objectives, having detailed knowledge of the facts that relate to the issues involved, and then applying legal principles to those facts in an effort to achieve my client's goals. Achieving my client's goals within the law based upon the facts of a particular situation is how I ultimately provide value to my clients.

Client Strategies

The most basic key to success in any legal situation is a positive client relationship. A positive client relationship in construction matters starts with the first client contact. A positive relationship is developed by being straightforward and genuine with the client. It is a disservice to the client to create unreasonable expectations or foster a client's belief that it will obtain a result greater than what is legally obtainable. A positive client relationship is perpetuated by being responsive to a client's needs and providing meaningful answers to their questions. I make sure clients understand that I am their advocate and that I am there to represent their best interests. I fully explain to them how any information they provide to me is protected by the attorney/client privilege and will be held in confidence. Because any information my clients convey to me will not be disclosed to any outside parties without their authorization, they can be completely open and honest in discussing the details of the case.

The information I strive to obtain upon initially meeting with a client is the basic facts about the issues in dispute in the client's own words. By allowing the client to tell the story, I get a better sense of the personalities and nuances surrounding the dispute. I review the contracts in question as well as relevant correspondence or e-mail with the client to understand how they interpreted the agreement and to know the meaning of their communications with others. This information helps me frame the facts and legal issues, and it allows me to begin formulating a strategy to address the issues and concerns of my client. Knowing and understanding this information allows me to provide specific advice to the client and identify other issues that may need to be addressed, and of which the client may not have been aware.

It is impossible to develop a meaningful and effective strategy without knowing the facts. It is a well-known law school adage that if the facts change, the law changes. In other words, a party's legal rights and remedies are directly affected by the factual circumstances surrounding the matters in dispute. It is important to know as much about the facts of the situation as possible in order to begin developing an effective strategy for resolving the dispute, whether the strategy is to vigorously pursue litigation or arbitration or to seek to settle the matter as quickly as possible through mediation or direct party-to-party negotiation.

Formulating Client Strategy

Once you have obtained the facts and information necessary to fully understand the matters in dispute, formulating a client strategy is dependant on several factors. The most important of these is the objectives of the client and whether those objectives are legally obtainable. To develop a client strategy in a construction matter, it is necessary to know whether the client's objective is to resolve the dispute as quickly as possible, to pursue all remedies available to enforce the client's legal rights, or some objective in between. A key factor in developing a client strategy is the economic cost and potential economic benefit of a particular course of action. Construction clients are businesspeople, and they are always interested in the bottom line financially. They understand there is a cost associated with the enforcement of legal rights, and it is important in developing a strategy to take the cost and benefit of a particular course of action into consideration before deciding on a particular strategy. Included in this analysis is an evaluation of the strengths and weaknesses of your client's legal position and your opposition's legal position. All of these factors impact your strategy, and they must be considered in developing your strategy.

There are five questions I routinely ask clients in developing a strategy for a construction matter:

1. What is in dispute?
2. What objectives do you hope to achieve in this dispute?
3. Within what timeframe do you hope to achieve your objectives?
4. What is the benefit to you in achieving those objectives?
5. What is the cost to you in achieving those objectives?

These questions are at the heart of every litigation strategy. To answer these questions, the client must focus on his or her goals and evaluate the cost and benefit of achieving those goals. In answering these questions, a client may realize that what were thought to be the goals in a particular situation are not. Based upon this evaluation process, new or different goals may be identified, which could change the strategy and decision-making process.

A client's financial status can impact the strategy taken in a particular case. For example, if a client is in financial dire straits and is the party from whom payment is sought in a particular construction dispute, this can be used to the client's advantage. If a client has no money or limited financial resources, this information can be used as a tool to resolve the dispute. Most parties seeking payment understand it is better to get some money from a financially distressed client rather than continuing litigation at the risk of receiving nothing from a bankrupt client. The client's financial status will also dictate the amount a client is able to spend on litigation, which affects whether the strategy taken is proactive or reactive.

Lastly, it is important to consider the non-legal ramifications of a client's construction matter. The client ultimately makes the decision on non-legal business-type issues that may impact a construction matter, and it is necessary to understand the ramifications of various business issues when developing a construction dispute strategy. Sometimes the non-legal concerns may be so overpowering that they impact the legal strategy. For example, if there is a great concern about important confidential or proprietary information being disclosed in a litigation matter, this may cause the client to work toward resolving the matter more quickly in order to alleviate the risk of disclosure of this very important business information.

Representing a Client

After obtaining all of the information possible through interviews of client representatives and review of relevant documentation, my ultimate approach to representing that client will depend on the client's strategy and objectives. If the client is seeking payment for labor, services, or materials provided, the strategy likely involves promptly communicating the client's claim to the opposing party either by telephone or through written correspondence. This can take the form of a demand in writing for the amount owed and notification that if the amount owed is not paid promptly, the client will pursue all legal remedies at its disposal. Conveying this message to the opposing party does not prohibit further negotiation in an effort to resolve the matter, but it does make clear the seriousness of the matter and that the client is committed to pursuing its position through litigation if necessary. Regardless of the dispute, it is always important to initiate communication with the opposing party. Your initial communication will set a tone for the negotiation. Early communication allows you to gather information regarding the opposing party's position and rationale, which may ultimately benefit you and your client in unexpected ways. I typically take this new information and evaluate it in light of the current client strategy, as any new information may cause the strategy to be altered.

If I am representing an owner or general contractor and the dispute is over paying a general contractor or subcontractor, my approach may be slightly different. Early communication with the other side may not be as high a priority as it would be if my client were seeking payment. Rather, I am more likely to wait until the party seeking payment makes contact with me or my client. When that occurs, I use the opportunity to gather more information regarding the opposition's claims and evaluate whether it would be beneficial to convey my client's position.

Whether you are representing a client that is seeking payment or a client from whom payment is sought can affect when and how you make the client's legal position known. The case strategy also dictates when, how, and whether I will communicate that my client is willing to consider any reasonable proposal for resolution. If my client desires to resolve the matter quickly, I will be more proactive in pursuing settlement negotiations in order to reach a satisfactory resolution.

Developing Case Theme

The theme of the case is developed over time based upon the facts and legal issues. In developing a case theme, I try to keep things simple by couching the theme in terms that are universally understood. A client can also be very helpful in developing the theme of the case. I like to ask a client how they felt when a particular situation arose or when they were treated a particular way by the opposing party. Frequently, when a client describes particular circumstances of a dispute, they will use phrases such as "we only wanted what was fair" or "we wanted [the opposing party] to do the right thing." Likewise, I try to develop theories based upon concepts of fairness and doing what is right. Themes based upon trust or honesty are good case theories. Obviously, there are times when the theme of the case is not easily explained, but if you do not have a theme in which you and your client believe, your effectiveness is significantly limited.

The Goal-Setting Process: Understanding Client Motivation

Understanding the client's motives is important in developing a successful case strategy. A client's motivation in any given case can be revealed by asking questions about the client's objectives and the costs and benefits of achieving those objectives. When a client is motivated by the belief that he or she has been treated unfairly or the opposing party has been dishonest, the client's primary focus may not be on the economic costs and benefits of a particular course of action. In short, the client may be angry with the opposing party and want to seek punishment regardless of the cost. Part of my role as counsel is to catch such situations early in order to make sure the client is evaluating the case in an objective manner rather than allowing emotions to impair good legal decisions.

A client's ultimate goals can often be identified simply by asking the client directly what those goals are. On occasion, a client's stated goals and their actions may be inconsistent. When an apparent inconsistency arises, it is necessary to bring this to the client's attention and work through any inconsistencies. By working through this process, the client's goals may change or its actions may change to become better aligned with the case strategy. Sometimes a client may not know what his or her ultimate goals are, and in this circumstance, it is important to explain to the client what the potential outcomes of a particular course of action may be. Helping a client understand the possible results can facilitate the client's definition of his or her goals. This in turn will help define the case strategy.

On the other hand, a client may have a strong idea of his or her goals, but those goals may be unrealistic. In this situation, it is important to tell the client as early as possible in the process that his or her goals are either unattainable or highly improbable. An important part of helping a client understand his or her case is to explain the legal process and what is required to bring a case to conclusion. I often find that clients are surprised at the amount of time it can take to bring a case to conclusion, which can be two to five years or more depending on the jurisdiction, the number of parties, and the number of issues in dispute. Walking the client through the legal process, from the filing of a complaint to the entry of a judgment to the possibility of appeal, helps a client get perspective on the dispute and what is involved in ultimately prevailing in a construction litigation matter.

The Attitude Effect

A client's attitude in a construction dispute can significantly impact the manner in which I proceed. Certainly, if a client does not want to pursue a vigorous defense or a vigorous prosecution of a construction dispute, we may take a quieter stance, which may be the best strategy depending on the facts of the case. Likewise, if a client wants to protect its interest at the lowest possible cost, the way the matter is handled will be impacted. Under this scenario, a reactive strategy may be followed and the representation will involve responsive action to the opposing party instead of proactive conduct. Once I have consulted with a client and we have agreed upon a case strategy, that strategy will be pursued until the circumstances change or we determine a different strategy is necessary to achieve the client's objectives.

If the client is unwilling to settle the case, that attitude will impact the case strategy. In this situation, the strategy may become more focused on preparing the dispute for trial or arbitration as expeditiously as possible. However, construction clients are generally willing to settle their disputes if the terms of settlement are reasonable. By and large, construction clients understand that negotiating the settlement of a legal dispute is similar to any other business negotiation. Construction clients are generally interested in minimizing risk and maximizing return on their investment, so it is an easy transition for most construction clients to use those concepts in a construction dispute.

It is unusual, but not unheard of, for a construction client to be ambivalent about his or her own construction dispute. However, when there is a sense of ambivalence, it does not change the options for proceeding in a particular matter. It may require that the options be explained in detail and on more of a continuing basis to ensure that the client understands the options and selects a particular strategy. It may also require a more frequent re-evaluation process to make sure the client's desired course of action is being pursued.

Dealing with "Bad Facts"

In my experience, the best way to deal with any "bad facts" from a client is to recognize them for what they are. It is a mistake to ignore bad facts. In certain circumstances, it is preferable to acknowledge bad facts but not overemphasize them. One should keep in mind that in every case there are good and bad facts on both sides. It is necessary to identify all bad facts but work toward developing good facts that, hopefully, will outweigh the bad.

There are several strategies for dealing with bad facts. One strategy is to acknowledge the negative facts openly and then explain them away by focusing on the reasoning and motivation of the persons responsible for them. This strategy is most effective if the reasons and motives are sincere. Another strategy is to make the negative facts appear insignificant or not determinative of the ultimate outcome of the case. These strategies can help reduce the impact of the negative facts on the goals and objectives of your client.

Winning Strategies: Working toward Your Client's Best Interest

There are times when a client wants to pursue a strategy that is not in its best interest. In these situations, it is necessary to make clear to the client how a particular course of action will negatively impact the matter and present all possible results of a particular course of action to the client. If a particular strategy is outright unethical, I simply will not pursue it. Should the client persist in pursuing such a strategy, I make clear to him or her the negative results that may occur. If, after being informed of the consequences, the client continues to pursue an unethical strategy, I will withdraw from the representation. If the strategy is not unethical but may only result in a negative outcome for the client, I will continue to work toward convincing the client to consider other alternatives.

Again, the most important aspect of developing and implementing a proper strategy is to clearly define the objectives in a particular construction matter and to communicate the procedures that will allow objectives to be achieved. It is important to have clear and continuous communication with the client in order to develop and implement a strategy in a construction matter, because most construction disputes are evolving in nature. Continually re-evaluating the objectives and the procedures to achieve the objectives is necessary to develop and implement a winning strategy. The negative consequences of developing and following an improper strategy are ultimately a negative result and an unsatisfied client. The best way to avoid this situation is to communicate with the client and fully advise them of the potential outcomes of the strategy and to ask the client if those outcomes are what the client desires. If the potential outcomes are not what the client desires, obviously a different strategy must be developed.

Some of the mistakes I see attorneys making in developing and implementing client strategy are not knowing the facts before choosing a strategy and getting into the mindset that one strategy fits all. Choosing a strategy without knowing the facts and not adjusting the strategy as new facts are discovered is, likewise, almost always fatal to a satisfactory result. Using either a "Rambo litigator" strategy or a "do the bare minimum" strategy on every case can have negative results. A one-size-fits-all strategy rarely works. To avoid this mistake, it is necessary for attorneys to discuss case strategy with their clients early and on a regular basis throughout the course of the matter. This is the best way to know the desires and motives of the client and to achieve the client's objectives.

Trent A. Gudgel is a shareholder with Hall, Estill, Hardwick, Gable, Golden & Nelson PC in Tulsa, Oklahoma. Mr. Gudgel received his B.B.A. in finance from the University of Oklahoma in 1986 and his M.B.A. from the same school in 1990. He attended the University of Oklahoma College of Law and received his J.D. in 1990, and was a member of the Order of Barristers.

Mr. Gudgel focuses his practice in the areas of construction litigation and advice, commercial and business litigation, and real estate litigation and advice. Mr. Gudgel represents general contractors, subcontractors, suppliers, architects, engineers, owners, and financial institutions in all aspects of construction and business litigation, arbitration, and mediation. He is a member of the Oklahoma Bar Association, the State Bar of Texas, and the American Bar Association. He is admitted to practice before all Oklahoma and Texas state and federal courts as well as the Fifth and Tenth Circuit Courts of Appeals.

Dedication: *I would like to dedicate this chapter to my family and thank them for their unwavering support. I would also like to thank Carolyn Rumsey for all her help.*

Efficiently Assessing and Addressing Risk

William W. Pollock

Partner

Cranfill, Sumner & Hartzog LLP

Attaining a Successful Outcome for Clients

Practice as a construction lawyer is an extremely varied area. Construction is a field in which there are many inherent risks both financially and with respect to personal injury and property damage. Construction lawyers are often consulted to help minimize or avoid these risks. The work I do as a construction lawyer can be divided into three areas. The first area pertains to the work done before a project begins, during which phase I give advice to clients on contract forms, contract negotiations, and bidding and licensing requirements. With respect to the second area, construction lawyers are occasionally called upon to give advice for disputes that arise during the course of performance of a project, which typically involve negotiations over changes to the scope of the work involved and the associated additional costs incurred. The third area in which a construction lawyer is involved, and the area in which I spend most of my time working with clients, relates to disputes arising after completion of the project. These disputes include claims arising out of performance of the construction contract, which include claims for additional money owed for change order requests and claims for errors that were purportedly performed in the work by contractors, architects, and engineers. Such disputes may also include claims by workers injured during the construction process or by owners or adjacent property owners claiming damage to other buildings or property as a result of the construction. Most of my work as a construction lawyer involves assisting clients through these disputes and helping them resolve these disputes through negotiation or litigation.

The areas where I feel my services have the most value for a client are in helping them assess the risks involved and in deciding how to best use their resources to address those risks. Having been through a number of trials and countless mediations and negotiations, I have learned through experience that a "win-at-all-costs" attitude is hardly ever the right stance to take. An important component of my role as a counselor is to convey that thought to a client. Even lawsuits that have no merit unfortunately cost money to resolve, and any lawsuit is a drain on a company's financial and time resources. My services have the most value to a client when I can accurately gauge for them the potential risks involved in any claim, both in terms of the ultimate exposure to the claimant and the time and resources necessary to obtain a resolution, and help them make an intelligent risk/benefit analysis as to how best to allocate their resources to address their claim.

The Different Components of Construction Law

Major construction-related issues with which I help my clients are claims arising out of alleged errors in the performance of construction work, catastrophic personal injuries or death claims arising from workers injured during a construction project, allegations of errors in the plans and designs for a construction project, and bond and lien claims for monies owed on construction projects. Most of my work involves litigation, arbitration, and mediation for these various types of disputes.

The primary areas of construction law involve contract negotiation and dispute resolution. There are some other smaller, ancillary areas such as professional licensing and employment issues, but in a broad sense, most of the work for construction lawyers involves either some aspect of the contract negotiation phase or some aspect of the resolution that arises out of performance of the construction project.

Regardless of whether a specific case involves contract negotiations or dispute resolution, my role will involve helping clients assess, analyze, and avoid the risk inherent in the construction process. On the contract negotiation side, I work with clients to ensure that their contracts adequately and thoroughly define the scope of each party's responsibility and that the risk of loss on any project is properly shifted to the party with the most ability to control that risk.

The vast majority of my work is in the area of dispute resolution. Most of our clients are not intimately familiar with the litigation process, and we make use of our collective experience in this area to advise them upon the rules and procedures to be followed in litigation. I also work with clients to analyze the possible outcomes and help them decide how to best address the claims presented.

Financial Implications of Construction Law: Adding Direct Value

The service I find provides the most value to clients is when they ask for advice prior to a dispute arising. A client can typically save money in the long run by spending money up front to properly analyze the contract documents and help assess the risks before a project begins, and by helping to ensure that there is proper protection and risk management in place. For example, by ensuring that proper insurance coverage for all subcontractors is in place and that the contractual language meets statutes governing risk transfer, we can ensure that, in the event that a claim should arise, the defense and liability for that claim would be borne by an insurance carrier or subcontractor rather than by the client itself.

Common Client Mistakes

The biggest mistake clients tend to make that gets them into trouble with respect to construction law is entering into contracts "on a handshake." While it is certainly important to have a high level of trust with all parties with whom you enter into contracts in a construction project, it is foolish for any client in any industry to enter into a potentially multimillion-dollar contract without ensuring that the contract fully sets forth each party's obligations to that contract. By taking the time to work on proper contracts before the project starts, clients could ultimately save themselves many headaches later, but unfortunately there are many situations where clients rush into things and use either form purchase orders or no contracts at all.

Strategies to Help Clients with Construction-Related Issues

In analyzing any situation for a client, I try to devote a considerable amount of time to thoroughly explaining the process and the risks involved to the client. By providing a client with enough information on the situation at hand, the client can ultimately make the most informed and intelligent decision and decide how to best spend resources.

In order to be successful with a client as a construction lawyer, you must be absolutely frank and honest with them on all your opinions. You cannot sugarcoat your opinions or simply tell the client what you think he or she wants to hear. Rather, if there is bad news, you need to give it directly to the client without any hesitation. You must tell the client how to deal with certain issues, and if they have significant problems, you need to let them know immediately rather than wait until the eve of trial.

Another key component to achieving success in this role is keeping an edge in this industry by staying abreast of current and emerging trends. In order to stay current in the field, I am constantly reading both trade and legal periodicals such as the materials published by the American Bar Association, the Defense Research Institute, and the Associated General Contractors. I also try to attend both client seminars and legal seminars related to construction law.

The Initial Meeting with a New Client: Preliminary Information and Research to Obtain

Prior to the first meeting with a client on a new construction matter, the first step is always to determine the nature and scope of the client's problem and why he or she is consulting with an attorney. On the initial contact with the client via telephone, I will spend time with them exploring some of the basic facts involved in the controversy and the issues that have caused them to seek legal advice. I will then ask them to send me some of the relevant documents prior to the meeting. I will also conduct research on the client and any opposing parties, as well as the project in question. This provides me with some basic background information and a general understanding of what we are facing so I am prepared before entering the initial meeting.

When a client comes to the initial meeting, I ask them to bring all relevant documentation if he or she has not yet done so. Of particular interest to the construction attorney are any contracts between the parties and any documents relating to purported problems on the construction project. I will also ask the client to provide me with any documents given to him or her by the opposing party or the opposing party's counsel. This information will help me further define the scope of the issues and the problem. It will also enable me to determine if there are any other issues that may arise, even though they may not yet have been noted by the client. It is extremely important for me to obtain all of the relevant contract documents. Hopefully, the contract documents will be sufficient to help outline the nature and scope of the obligation of various parties to the dispute. At a minimum, it will help me determine what issues are not outlined by the contract documents. I also need to obtain any documents that relate to the problems at issue, as well as documents generated contemporaneously with the construction such as correspondence, project logs, and meeting minutes. At this stage, I do not need all documentation generated by the project, as sometimes that amount of material can be quite voluminous. Rather, I need the minimum amount of information necessary to allow me to fully understand the scope of the project, the role of each party, and the nature of the dispute. Hopefully, this information and documentation will also provide a good indication of the size of the problem involved and help me start formulating ideas as to the best resources to allocate to address this problem.

Determining the Best Strategy for the Client's Construction Issue

The factors I use to formulate a client strategy in a construction matter are governed largely by the size and complexity of the dispute. For example, in rather small or straightforward matters, the costs involved for legal advice may exceed the amounts at issue. In such situations, I would give the client some general advice on how to best address the situation and provide them with the information and tools needed to handle the matter on their own, which may be the best use of client resources in that particular case. Conversely, if there is a very complex or costly claim that would justify the allocation of significant resources, I would then address the strategy in a more thorough and aggressive manner and assume a high level of involvement in the process.

In asking questions of my client to determine the best strategy, my first questions revolve around learning about the client and their business. By asking probing questions about the client and their business, I try to gain an understanding of my client's needs and goals and how they can best be met. Secondly, I attempt to learn the same facts about the opposing party so I can understand the motivation behind that party and the dispute at hand. The third area in which I ask questions revolves around the project itself, with the intent of gaining an understanding of the nature and scope of the project that has given rise to the dispute. Fourth, I ask a number of questions about the dispute so I can determine the full scope of the issues between the parties. Finally, I always ask clients how they want to have the matter resolved or what their ideal outcome of the matter would be. This helps me determine where my client thinks the case needs to go and lets me know if I need to educate the client about additional options that may represent possible desirable outcomes.

General Approach to Representing a Client in a Construction Dispute

My general approach in representing a client involved in a construction dispute is to look for the most efficient way to resolve the dispute. While few lawsuits in general end up in trial, it is even rarer for a construction dispute to end up in front of a judge or jury. Further complicating this fact is that many construction disputes have complex legal and factual issues that entail significant risks for all parties if they are indeed to be decided by lay judges or juries. For those reasons, while it is sometimes necessary to go through significant discovery portions of the legal process, I have come to realize that a negotiated settlement of any dispute is typically the best result for a client. Depending on the personality and issues involved, it may be necessary to push a matter closer to trial in certain situations, but in all situations I am always looking for the most efficient and cost-effective way to resolve the litigation for the client. While the specific steps I might take to prosecute a claim differ from those I would take to defend a claim, there is always one constant regardless of the nature of the case: I will always strive for the same overall goal for my client, which is to achieve the most cost-effective and efficient resolution, irrespective of what side they are on in a construction dispute.

Establishing a Positive Client Relationship

The most important aspect of developing and implementing a proper strategy is obtaining the full trust and confidence of your client. By working with clients to let them know you understand and care about the best resolution of their problems, and that you will give them your accurate and honest assessment of a dispute, clients will hopefully place complete trust and confidence in your abilities and strategy and allow you to direct a course of any settlement or litigation.

In order to attain this goal of establishing a positive client relationship in a construction matter, I consider and analyze the issues at hand, after which I try to give my client an early and honest assessment of the dispute. At this point, I outline the potential risks to the client if the matter ultimately goes to trial. I try to explain both sides of the dispute to the client so the client understands the risks going forward. I also try to let them know steps involved and the costs necessary to take the matter through trial so we can determine the best use of their resources in resolving the dispute. I find that this level of frankness and directness helps put the client at ease and establishes trust between us. I often tell clients that, as their advocate, I can give a certain "spin" to the facts, and I will certainly do so if necessary at a hearing or trial. However, from an objective viewpoint, I may see the facts somewhat differently than how they view them or how I ultimately present them. By being up front and honest with my client, I think I can both temper their expectations toward the matter and give them facts they need to make a fully informed and proper decision. Given that all the advice I would give them is confidential, I feel I can speak openly and frankly to them about the dispute at hand. At that point in time, I will know whether the client is one who actually wants to listen to and follow my advice. If this is the case, both the client and I can decide if we want to go forward with my representation of them at this dispute.

From a professional standpoint, I do not let a client's attitude impact the veracity with which I proceed. I have found that my professional reputation is worth more than the benefit to be gained from any one client. Thus, if the client wants to take an inappropriate strategy on a case, I would advise them that it is against my best judgment and that I will not proceed in a manner I believe to be unprofessional and unproductive.

Further, if a client has an unrealistic goal, I try to help them understand the risks, costs, and benefits of reaching those goals. I have found that after some calm thought and reflection, most clients realize they have to set aside any personal feelings they may have developed and make what is the best economic decision for their company. The best strategy in helping clients understand the reality of the case is a straightforward and honest assessment of the facts and risks, as outlined above. These strategies are helpful in resolving construction disputes because the client gains a good feel for the attorney's judgment and understanding of the situation, and most clients appreciate an honest assessment of their issues.

It is also important for me to understand my clients' motives. There is no single item of information that enables me to learn a specific client's motivation. Rather, it is a judgment decision based upon a full understanding of the facts involved in a dispute along with an analysis of the risks and benefits associated with a particular case. Often, a client does not know what his or her own goal is, and by helping the client explore various options I can then best learn about his or her motivation in trying to resolve a dispute. Ultimately, unless you are representing an individual homeowner, the motivation in most construction disputes comes down to the best economic decision for the company, particularly when you get clients to focus on the risks and benefits and have them step away from any emotional ties to decisions that have been made.

Developing the Theory of the Case

In developing my theory in a construction dispute, I first attempt to define the problems at hand and then apply standard construction law legal theories to that dispute. The concepts of joint and several liability and tort damages versus contract damages are almost always involved in a construction dispute, as are concepts of risk transfer. The client's involvement in this dispute is critical, in that they help me understand the facts and nature of the dispute, and I then tailor the legal theories to those facts. I find that most clients let the attorney decide the legal theories as long as you have the facts correct in developing that theory.

Settling Construction Matters: The Client's Willingness to Settle

I have found that in most construction disputes, while it may take some time for the client to fully understand the economic realities of a case, most clients are ultimately willing to settle a dispute rather than risk taking the matter to trial. I believe clients in construction-related matters are more likely willing to settle a case than clients involved in other forms of litigation, particularly when it involves complex factual issues relating to the construction project itself. Given that juries and judges are not typically well versed in construction practices, and construction disputes often involve technical terms and concepts, the parties usually believe it is better to work out a dispute themselves rather than let an unsophisticated jury determine that dispute. A maxim that has been stated with respect to such scenarios is that, "A good settlement is when all parties walk away equally unhappy." I try to convey this to my clients in all situations so they may realize that settlement is the ultimate control over a dispute, and by letting this dispute go to a judge or jury they lose control. The only situation in which I find this does not hold true relates to personal injuries occurring for construction site workers. While I am always representing owners or contractors in such disputes and do not represent injured workers, I find that their motivation, while ultimately economically based, is often mired with personal issues, particularly when you are dealing with a less-educated or sophisticated plaintiff.

If I am given "bad facts" by a client in connection with his or her construction dispute, my role is to determine ways to keep those facts from being discovered or to think of alternative theories by which to counter the impact of those facts on a dispute. Ultimately, however, if there is simply no way around those facts, I advise the client that we must meet them head-on and that this will impact the settlement value of their case.

The Impact of a Client's Financial Status on Strategy

A client's financial status always impacts the strategy taken toward a case. If a client has limited resources, it is often best to approach opposing counsel and determine if the case can be settled early on in litigation. By explaining to opposing counsel that there are limited resources that might go toward any settlement, coupled with the fact that litigation costs will likely eat up those resources, I find this makes an early resolution of a case easier to achieve as long as the situation is thoroughly and honestly explained to opposing counsel and supported with strong documentation.

Considering Alternative Measures

Alternative measures are always important to consider, and both client and counsel should always remain open to alternative ideas. Strategies typically change during the course of a construction dispute, and if you are inflexible and tied only to one solution, ultimately you are not working in your client's best interest.

To fully advise a client, a construction lawyer must understand both the legal and non-legal ramifications of any dispute. For example, there are business relationships that need to be salvaged out of the dispute, which is an important factor that will affect how you develop your legal strategy in a particular suit. For example, a client might pick up a portion of a settlement for which it was not otherwise legally obligated in order to keep a very important customer happy. In such a situation, the client's overall financial status takes precedence over who could prevail in a court of law.

If a client wants to have me follow strategy that is either not in the client's best interest or unethical, I will explain to the client the negative consequences of that action and advise them that it is not in their best interest. If it involves unethical conduct, I tell the client without hesitation that we will not follow such a course of action. If the client insists on such a course of action, he or she will need to find new counsel. Fortunately, I find that such situations are extremely rare and that in practically every situation, when you have developed a good relationship with clients, they will trust your advice as their attorney on strategy.

Common Mistakes Attorneys Make with Regard to Developing and Implementing a Client Strategy

The biggest mistake I see attorneys making is allowing the client's emotions to control the outcome of the dispute or strategy taken. There are situations in which clients are upset over a construction dispute. Sometimes these emotions are justifiable, and sometimes they are less so. Regardless, the best course of action almost invariably is to set aside those emotions and deal with the case on the basis of a cost/benefit analysis. When attorneys allow a client to improperly inject those feelings into what is essentially a business dispute, it fails to serve anybody's best interest and suddenly serves to increase the costs and animosity between the parties.

William W. Pollock is co-chair of the firm's liability litigation section and concentrates his practice on construction law, environmental and toxic torts, product liability matters, and other complex litigation. He has been a member of Cranfill, Sumner & Hartzog since 1992. Prior to joining Cranfill, Sumner & Hartzog, he practiced with a large firm in Washington, D.C., for five years after graduating from law school.

As lead counsel, Mr. Pollock has tried more than thirty cases before a jury, with favorable results in more than 90 percent of those cases. He has handled matters in North Carolina Superior Court in most of the counties in North Carolina, and before all federal district courts in North Carolina, as well as before the North Carolina Court of Appeals and Supreme Court.

Mr. Pollock's representative clients include general contractors, architects, engineers, product and chemical manufacturers, and insurance and surety companies. He serves as the firm's primary contact with the Harmonie Group, a national membership organization of independent law firms that provides clients with access to defense firms and attorneys that handle complex and difficult high-stakes litigation throughout the United States.

Dedication: *I would like to dedicate this chapter to my wife, Maggie, who has supported me without fail through the long hours and hard work throughout my career. I would also like to thank my invaluable paralegal and legal assistant, Jane Johnson, and Kim Vilanova for their assistance in preparing this manuscript.*

Getting Involved Early and Making a Difference

Craig A. Ramseyer

Partner

Procopio, Cory, Hargreaves & Savitch LLP

Construction Law

The areas of construction law handled by my firm include contract drafting and review, False Claims Act defense, bid protests, lien and payment bond claims, schedule analysis and productivity analysis, and work scope issues.

- *Contract drafting:* We draft tailor-made subcontracts; we review, modify, and negotiate standard form agreements such as American Institute of Architects contracts; and we consult with clients on the ramifications of particular contractual provisions.
- *False claims:* We defend federal and state false claims allegations brought by public entities.
- *Bid protests:* We analyze whether our client has grounds to challenge an award to the apparent low bidder or, conversely, whether a bid protest by a disappointed bidder should be denied.
- *Liens and payment bonds:* We record liens and make payment bond claims on behalf of clients, and we defend the same.
- *Schedule analysis:* We retain scheduling experts to analyze construction schedules in the prosecution or defense of delay claims by others.
- *Productivity analysis:* We retain consultants to analyze productivity and loss of efficiency claims in the prosecution or defense of an action.

We believe we can add value for our clients by consulting and advising regarding claims before litigation is filed, before attitudes harden, and before relationships are irretrievably broken. Once the claims have been analyzed and the client has been advised of the analysis, we can foster negotiations that hopefully will result in resolution of the issues prior to litigation.

In our practice, we advise clients on dispute avoidance, advise clients on construction issues that arise on projects, and defend and litigate construction claims. We would rather assist a client through a difficult project, with the result being a completed project without litigation, than be involved in a project through expensive and time-consuming litigation because we are retained too late in the process to head off litigation.

The major construction-related issues we help our clients with include delay claims, productivity claims, scope of work disputes, payment disputes, and professional liability of architects and engineers. We provide value for clients by emphasizing counsel during the construction process to avoid and resolve disputes before they mature into litigation.

Areas where clients seem to get themselves into trouble include the failure to properly document a project, including full and complete daily reports of contemporaneous job site activities, and the failure to provide notice of claims in accordance with contractual requirements.

In our experience, it is not uncommon for contractors' files to contain less than complete and accurate daily reports. Everyday demands of a superintendent, whose responsibility it primarily is to prepare daily reports, often leave little time for recitation of all the facts and circumstances occurring on the job that may prove important at a later point in time. While our retention does not generally extend to educating experienced superintendents on how to fill out daily reports, we do emphasize whenever the opportunity presents itself that the daily reports need to be full, complete, and accurate in order to provide the full support necessary for the defense or prosecution of claims.

To assist our clients in construction-related issues, we emphasize timely response to phone calls, e-mails, and questions the client may have, honest evaluation of risk and reward with regard to advancing a particular position, and a clear plan for the case in the event an early resolution cannot be obtained. Once litigation is inevitable, we try to provide aggressive representation to ensure that our opponents are responding to our initiative, rather than the opposite.

We stay on top of our knowledge and keep our edge by being active in the American Bar Association, reading construction periodicals, and discussing current construction topics arising in the industry with our clients. In particular, the American Bar Association's Forum on the Construction Industry puts out a publication titled *The Construction Lawyer*, which contains quarterly articles of interest to construction practitioners. The forum also sponsors two conferences a year dealing with particular and topical issues for construction practitioners. Finally, we receive on a daily basis the advance sheets of cases in our jurisdiction that have published decisions. The same are reviewed for procedural and substantive issues that may be of benefit to our practice and our clients. We believe our experience and expertise in this particular facet of the law, and accurate assessments of risk and reward, contribute to success in representing and advising our clients. Success can be measured in monetary terms or in reaching a business decision for the clients in a cost-effective manner. In many instances, a quick resolution may prove in the long run to be more economically beneficial than expensive and time-consuming litigation that may yield more money but at a higher cost.

The Client/Lawyer Relationship

A positive client relationship can be fostered by establishing a construction budget and plan for the construction, and thereafter updating it so the client is always aware of the expense and future expense of the litigation. It is also important to share all pleadings and important information and involve the client in review of the same. Regular contact lets the client know you are giving his or her case priority and prevents any surprises as to the future handling of the case. The main benefits of establishing this rapport are that the client should be satisfied that it has the correct attorney to handle its matter, the client is knowledgeable about the status and thus is less stressed about what is going on in its case, and counsel can continually reassess risk and settlement strategy. This practice also provides insurance against malpractice claims and increases the odds that you will be able to retain that client for future matters.

The attorney should never lose sight of the fact that despite establishing a personal relationship with the client, the client is fully aware he or she is compensating the lawyer at a considerable hourly rate. The lawyer should not mistake mutual respect as a sign that he or she can slack off or fail to keep the client informed and updated.

Part of the relationship is involved in learning and understanding the client's motivation during a dispute or other legal matter. The client's motivation in the construction dispute is usually economic. Lawyers in construction disputes are generally redistributing money from one entity to another. In other words, funds are redistributed from a public entity to a contractor in the event the contractor has a valid claim, and from a contractor to a subcontractor in the event the subcontractor has a valid claim against the general contractor. In like fashion, funds are redistributed from an insurer to a contractor or owner in the event there is a covered loss. Thus, as noted, most construction disputes are economic in nature. But if the motivation in a particular matter is other than economic, you will learn how difficult working with the client will be, and you will be able to assess how difficult it will be to settle or otherwise resolve the case.

One way to learn clients' ultimate goals in a construction matter is simply by asking them. What do you consider a successful result to be and/or what do you hope to achieve at the conclusion of the representation? It could be money or it could be maintenance of a damaged relationship. In terms of strategies to help clients understand the reality of a case, a lawyer must be honest and tell them the strengths and weaknesses of each case. Straightforward advice is what you are being paid for, and it is helpful in any construction dispute or litigation matter of any kind.

We try not to allow the client's attitude to impact the aggressiveness with which we pursue any case. Our goal is to pursue each case aggressively. We will not become overly aggressive simply because the client wishes us to take that tact. If the client is unhappy with our representation, the client is free to go elsewhere. An overly aggressive attitude is generally not conducive to achieving the goals of the client. Lawyers sell time, and successful lawyers can and do usually allocate their time to more reasonable clients than to those who are difficult. Furthermore, many clients mistake aggression and animosity for effective representation. In those instances, we have advised clients we chose to no longer continue our representation.

I do not often find clients ambivalent toward their construction disputes. However, to the extent they are, they vest more discretion in me as counsel. Therefore, they are more open to my recommendations to resolve the case. As for clarifying the possibilities to ensure the course the client actually wants is followed, it is up to the lawyer to have discussions up front and then regularly thereafter with the client to make sure each are on the same page and are proceeding down the agreed-upon path.

If clients want to follow a strategy that is unethical, we will not represent them. In some instances, we will weigh out alternatives and clients may pick an alternative we feel is not in their best interest or may have other negative consequences. Under these circumstances, it depends on the clients as to whether we proceed with the matter or ask them to find other counsel. If we perceive the path they wish to pursue is not in their best interest or may have other negative consequences, it is our job to fully advise them and allow them to weigh the risks and benefits in proceeding in such a manner.

We do not necessarily set boundaries with our clients. We solicit their input as to the relative value of potential deponents, we provide discovery plans they can review and comment upon, and we remain open to discussing any strategy we are pursuing in the case. While we do not allow them to dictate the arguments that may be advanced in any particular motion, allowing the clients the opportunity to be involved in the prosecution of the case helps ensure their satisfaction with the work you are performing on their behalf.

Preparation

Our general approach in representing a client is to first obtain contemporaneous client documents and basic information from the client. These documents include things such as daily reports, schedules, contracts and subcontracts, correspondence, internal memoranda, and electronic documentation such as e-mail, payment applications, and accounting information. Once the documents are obtained, we prioritize depositions and notice the most important depositions. Once the most important depositions have been taken, we suggest going to mediation before completion of the remaining discovery. We take this approach and find it helpful because it is always important to discuss resolution of the case short of a full-blown trial. The parties at the outset of the case are often overly confident of their position and will not change their negotiation position until they have incurred financial pain in attorneys' fees or attended depositions or been deposed, and thereafter realize their position may not be as strong as they previously thought. If mediation is scheduled after the higher-priority discovery, both sides will have learned more about the case so they can adjust their perceptions. This approach generally does not vary according to the client or type of dispute.

Prior to the first meeting, we want to check for conflicts, because we want to avoid any embarrassing situation where we could find ourselves adverse to a client or client-related party. We also review the client's Web site, as well as the Web site of any opposing party or counsel, so we are knowledgeable about the business of our clients, their current activities, and the same with regard to our potential opponents. We also check with the industry regarding the client's standing within that industry, rumor on the street regarding a particular project, and the client's past success or failure with similar types of projects. All this information provides us with background to be knowledgeable in our initial consultations with clients and to begin to assess the merits of the client's position.

When we first meet with a client, we will solicit the specifics of a dispute. In particular, we are interested in the terms of the contract; the location, organization, and quality of contemporaneous project documents maintained by the client; the background experience of key client personnel; and an accounting of the monies due or claimed due on the contract. This information is necessary to get a full picture of the dispute so we can ascertain the appropriate strategy for the client. Depending on the responses to these questions, we may find it more appropriate to pursue an early resolution of disputes or we may learn it is likely the parties are hardened in their positions, and thus difficult litigation will ensue.

The types of questions we ask include:

- What is the accounting on the contract?
- What are the causes of delay as you see them?
- What is the quality of your project documentation?
- What is your opponent's response to your position on the disputed issues?
- How would you characterize the relationship between the parties in light of these disputes?

The answers to these questions will help us determine how far apart the client and opponent are, how much emotion is in the case and preventing a business resolution of the disputes, how easy or difficult it would be to find supporting evidence, and how much money is in dispute. It further allows us to analyze the reasonableness of our client's position versus the position of the opponent.

The documentation we feel is important to receive includes daily reports, change orders and proposed change orders, payment applications and billings, e-mail, personal files maintained by directly involved individuals, correspondence files, and construction schedules.

The theory of the case is based on what we believe will resonate with the fact finder. I generally do not involve the client in this formulation, because I am being paid to quarterback the matter and I am using my judgment and experience to determine what themes will work. The theory of the case involves not only the justification for why your client is entitled to judgment, but also a theme. The theme could be as simple as that running a business is equivalent to pedaling a bicycle, and one has to keep peddling in order to avoid collapse. Therefore, the theme will be that the plaintiffs simply stopped peddling. The theme, in theory, would be one the fact finders could understand in the context of their own lives.

Every case has "bad facts." If the facts are too much in favor of one side or the other, there is not much of a dispute. I tell the clients the facts are what they are, and our job is to argue why those facts are not as important as the other side believes. When you have bad facts, you want to be sure they come out in your case rather than allowing your opponent to expose them.

The strategy to minimize the impact of negative facts is to not run from them. Rather, bring them out in opening statements and provide an explanation rather than allowing them to be a surprise during the course of the trial. It is important to deal honestly with the negative facts and advance plausible explanations as to why those facts should not sway the jury or fact finder from still finding in favor of your client. This strategy is beneficial to the client in that it appears the client is being open about the strengths and weaknesses of the case and the lawyer is straightforward and honest.

Non-legal ramifications on a construction matter are important. The client may want to avoid publicity over a delayed project or avoid being tagged as a litigious entity. The client may be seeking other work from other entities and choose not to have pending litigation, which will count against it in consideration for the other work. All these considerations are important and should be ascertained at the outset of the case. Any strategy must take into account the non-legal ramifications so the client is best able to achieve satisfactory results. Under any circumstances, the satisfactory result may not always be the highest dollar amount that can possibly be obtained through trial or settlement for a lesser amount.

Strategy

The most important aspect in developing and implementing a proper strategy in a construction matter is to understand the client's needs and objectives. Their ability to finance the litigation, the costs in proceeding, and the result they wish to achieve all must be considered in developing a proper strategy.

The negative consequences of following an improper strategy include unnecessary expense and an unhappy client. If the strategy is not designed to get the client the results needed, you will spend the client's money moving the case from side to side rather than forward. If you are on the path to a result you believe is in the best interest of the client but does not satisfy the client's motives or desires, you will not represent that client for long.

Consequently, full and open discussion of the goals and objectives for the litigation must take place at the outset of the case. However, if negative consequences are encountered, you must sit down with the client and retool your strategy to ensure that you and your client are on the same path to the desired result.

In formulating a strategy, we first attempt to ascertain the client's objectives. For some clients, resolution in and of itself is important in order to preserve a client relationship. In other instances, financial considerations predominate, as it is important for the client to recover or reduce its loss. The factors impacting our strategy include the size of the dispute, the financial strength of the client, the venue of the case, and the presence or absence of a fee provision. All these factors allow us to analyze the client's resolve and financial ability to pursue its position aggressively. If we are on the defense side, we want to be sure there are no undisputed amounts that may be owed to the claimant such that even if we otherwise win on the disputed issues, the claimant will remain the prevailing party and get fees according to a fee provision.

Litigation is expensive, and therefore the better able the client is to handle the costs of litigation, the better his or her chances of having his or her case fully prepared. Often, however, smaller clients have financial problems as a result of the dispute and do not have the ability to keep up with fees as they are incurred. The decision to continue representation of clients in this scenario will depend on the existence of a prior satisfactory relationship, the client's sincerity and effort in staying as current as possible, security that may be available, and the anticipated fees and costs still to be incurred. An example of this would be a recent matter in which we successfully defended an alter ego claim against the officers and directors of a construction company that went bankrupt. The officers and directors had infused the company before its failure with significant personal funds and were without the cash flow to manage the expenses of litigation. Accordingly, we accepted a trust deed on a piece of property that, when sold, had sufficient equity to satisfy the accrued fees.

The client's willingness to settle, or lack thereof, certainly impacts your strategy. If the client wants to go to trial and not settle, we can focus our energies on getting ready for trial as expeditiously and cost-effectively as possible. The problem comes when a client is overeager to settle the case. With such a client, you are often restricted in the work you are able to perform, and this in turn makes you less effective. Cases settle from positions of strength, not of weakness. Therefore, if you have developed a weak position because the client has restricted your ability to prepare your case, the case will not settle or it will settle for less than its full value. I do not believe construction industry clients are more or less willing to settle than other industry clients.

I am always open to alternatives as to new strategies for construction disputes. Those alternatives may be a creative manner of settlement in which there would be less financial consideration incurred but perhaps an agreement to do future work together. Many cases get stuck because the parties do not think creatively in working toward an appropriate settlement.

The biggest mistake I see lawyers make is that they do not develop and implement a client strategy. Many lawyers simply take certain steps without really thinking about what they hope to achieve. The consequences are unnecessary and wasteful work resulting in no benefit to the client. I would advise all lawyers at the outset of litigation to understand the client's goals and objectives, understand their financial ability to prosecute the case, and discuss with them alternative remedies that might prove satisfactory.

In closing, I note that the vast majority of construction cases settle. The fact that they settle suggests the resolution is at least satisfactory to both sides. As noted above, that resolution can be purely financial or it can involve more creative aspects. For instance, in prior cases as part of settlements, we have agreed to provide a first look at future construction work to the party with whom we are settling. In other instances, we have settled after we have agreed to provide access to experts and our own clients for use in the defending party's future indemnity claim against third parties. Other settlements have been reached with agreements to a cooling off period in which one or more parties agree not to bid on the other party's work for a stated period of time.

Craig Ramseyer is a Partner at Procopio, Cory, Hargreaves & Savitch and is the leader of the firm's litigation team. His practice emphasizes construction law and litigation. He is experienced in handling disputes arising on both public and private works of improvement, inclusive of contractual disputes, bid protests, mechanics' liens, stop notice and payment bond claims, workout agreements, design professional errors and omissions, and delay, disruption, and inefficiency claims. Mr. Ramseyer has represented general contractors, subcontractors, public and private owners, and sureties in various matters over the course of his career.

Mr. Ramseyer received his undergraduate education from Miami University in Oxford, Ohio, and earned his law degree from the University of San Diego in 1981, where he was a notes and comments editor for the San Diego Law Review. *He was admitted to practice in the U.S. District Court as well as the Central and Southern Districts of California in 1981. He was later admitted to practice in the U.S. Court of Appeals for the Ninth Circuit, the U.S. Court of Federal Claims, and the U.S. Supreme Court. In June of 1993, he was the principal founder of the law firm of Ramseyer & Kuhlman APLC, where he worked until Ramseyer & Kuhlman merged with Procopio in May of*

2001. Mr. Ramseyer serves on the panel of arbitrators for the American Arbitration Association. He is also a member of the State Bar of California and the American Bar Association Forum on the Construction Industry. Most recently, he was named to the 2007 edition of The Best Lawyers in America.

Prepare, Pay Attention to Details, and Communicate

Joseph C. Kovars

Principal

Ober, Kaler, Grimes & Shriver

My Practice

For more than twenty-five years, I have represented clients across a broad spectrum of construction, contracting, and procurement activities. My clients are owners, contractors, subcontractors, and sureties. The disputes often entail multiple parties and complex projects. I have successfully handled delay claims, changed work claims, defective specifications, contract terminations, bid protests, bond claims, mechanics' liens, defective work claims, and injunctive actions. My practice takes me before many federal and state courts, boards of contract appeals, the American Arbitration Association, and the comptroller general. I have also been an arbitrator with the American Arbitration Association for more than fifteen years.

Types of Cases

Delay Issues

The truism "time is money" has no greater application than in the construction industry. When a project is delayed, it is inevitable that additional costs will be incurred by the parties. It is not unusual for each side to blame the other for project delays and to seek reimbursement of damages from the other party. Many of my cases involve prosecution or defense of delay claims. My work includes assisting in the preparation or critique of the delay claim, helping prepare and assemble documentation in support of the client's position, analyzing project records to identify and establish delays, evaluating cost information, and then assisting with the dispute resolution.

Changes/Disputed Extra Work

Virtually all construction projects involve changes in the scope of work. At times, the parties disagree over the amount of the adjustment in contract price or time due for the change. Other times, they disagree over whether a change has occurred at all. For example, the owner's architect may interpret a contract requirement in a way the contractor believes represents a change in the work. The architect, however, may believe he or she is simply enforcing the contract. The contractor may submit a claim for a "constructive change." I counsel my client in evaluating the merits of the dispute and assist in the dispute resolution.

Contract Terminations

At some point in the project, one party may decide he or she wants a "divorce" from the other party. There are two main types of contract terminations: a "termination for default" (that is, for cause) and a "termination for convenience" (that is, a no-fault termination). Not surprisingly, the one who terminates the contract for cause often suffers damages in completing the work. The one whose contract is terminated may suffer damages as well, including lost profits or business opportunity costs. The terms of the contract often establish the substantive and procedural requirements for either type of termination. A wrongful termination may result in liability for the one who pulls the trigger. I counsel clients concerning whether termination is justified and appropriate, how to calculate damages, and in dispute resolution.

Bid Protests

In the public procurement arena, public owners must, by statute or regulation, treat bidders fairly. Often, they are required to award the contract to the lowest bid submitted by a responsible bidder. There is usually a procedure available for a disappointed bidder to challenge the procurement decision. I advise clients, who are either the disappointed bidder or the successful bidder, concerning the bid protest, and I litigate resolution of the protest for my clients.

Preparing the Case

Regardless of the case, the thought process works in a similar fashion. I start by gathering and analyzing the facts. I review the contract. I consider the applicable law and research the issues I am not familiar with. I challenge my client's position. I evaluate potential witnesses and consider the likely forum for resolution of the dispute. I then engage in a cost/benefit analysis where I consider the size and strength of the case and the risks of litigation against the possible litigation expense. At that point, I make recommendations to the client and together we agree upon a plan of attack.

As more information is obtained, whether informally from the client or formally through the discovery process, I reconsider the cost/benefit equation and advise the client if I believe we need to adjust our strategy.

In all cases, I try to think of the quickest and least expensive route to resolve the dispute. My approach is to be prepared, be relatively quick to implement the strategy, and be honest with my client regarding my appraisal of the dispute. When I represent the plaintiff, the goal is to try to recover the monies due as soon as possible. This may be accomplished by a settlement meeting or a structured settlement process such as mediation. One has to time the settlement, however, to maximize the likelihood of success. This may require taking limited discovery from the other side to educate it concerning the weak aspects of its case and why settlement is just as much in its best interest as it is in my client's. Always, one must keep in mind the intended audience. A presentation to a sophisticated opponent will differ from one to a judge who seldom hears construction cases. We will talk more about this later.

When I represent the defendant, I still try the quick route to resolution, because litigation costs can color, as they should, the client's view of a successful outcome. A sizable recovery can be eaten up by litigation costs, resulting in an unhappy client. To me, it is often more cost-effective to tie down key facts before the other side is prepared. Generally, I do not believe a defendant should take a "rope-a-dope" approach with as little activity as possible until a court-ordered settlement meeting or trial.

The quick route is not always an early settlement meeting. I had a case recently where, representing the defendant, I found some damaging documents amongst the plaintiff's files. Rather than discuss the documents in a settlement meeting, I took a key deposition and forced the principal to make admissions that greatly hurt his case. A day after the deposition, the plaintiff's attorney called to tell me he was dropping the case. My client paid *nothing* to settle the case.

If both sides have claims to make against the other, I race to the courthouse to file my case first. I prefer representing the plaintiff rather than the defendant/counterclaimant. The judge, jury, or arbitrator generally perceives the plaintiff as the wronged party, while the defendant's claims are sometimes discounted as mere "makeweights." The initial perception, of course, may change during the course of a trial, but I prefer starting with the advantage of representing the innocent, wronged party seeking recovery of his or her damages.

Documentation

The type of dispute dictates the required documentation. The dispute that requires consideration of the most project files is the delay claim. Almost all project records are fair game for proving or disproving the claim. That is because, in considering a delay claim, the *cause* of the project delay is often the heart of the dispute. Frequently, each side claims the delay was caused by things that were not their responsibility. If two critical delays are present, one due to a compensable cause and the other due to a non-compensable cause, the delays are considered "concurrent" and neither party may recover damages for the other.[1] So delays from any and all sources must be considered and determined whether they caused an overall project delay (a "critical delay").

Important documentation includes:

The Contract Documents

The contract usually defines which delay events are "compensable" (money and time), "excusable" (time only), or "non-excusable." The contract may also have procedural requirements for perfecting claims such as notice and claim backup deadlines.[2] In addition, the contract may have a liquidated damages clause, usually providing a daily rate to compensate the owner for delays by the contractor. The daily rate substitutes for actual delay damages by the owner (regardless of whether actual damages are more or less than the liquidated damages). Finally, the contract may have limits on the types or amounts of delay damages recoverable or may prohibit the recovery of damages for delay altogether.

---

[1] *William F. Klingensmith Inc. v. United States*, 731 F.2d 805 (Fed. Cir. 1984) (no recovery absent clear apportionment of delay attributable to each party)

[2] See Appendix G.

Project Correspondence

The correspondence may indicate the delays and may constitute evidence of the notice required by the contract. E-mails are an increasingly important component of this category.

Electronic Project Schedules and Updates

Most schedules use the critical path method of scheduling. To be compensable, the delay must be "critical." To ascertain if the delay is critical, one must analyze the project schedules. It is important to understand how the project schedule was updated, including whether activities were added or deleted, whether logic connections between activities were changed, or whether durations of activities were modified. This requires examination of the electronic schedules, not just the hard copies, because the hard copies often do not show this information. Software programs like Claims Digger quickly identify the changes from schedule to schedule.

Progress Narratives

Many contracts require the contractor to provide a narrative with each schedule update. The narrative is supposed to identify any schedule slippage, the cause, and how the contractor will address the problem.[3]

Progress Meeting Minutes

The minutes document discussions of job issues held at regular progress meetings. Delays sometimes may be identified and quantified through information in the minutes.

Contractor's Daily Reports

The daily reports may help measure performance of certain activities and may indicate causes of delay. They may also be useful in quantifying the delay and impact costs such as extra labor or equipment expenses.

---

[3] See Appendix H.

Inspector's Reports

The reports of the inspectors for the owner or code authority may provide useful information on delays and when activities were completed (because they passed inspection).

Time Extension Requests

The contractor's requests for additional time obviously are relevant to delay analysis. A contractor who submits an excusable delay request, such as for extraordinary weather, normally can't claim the same period of delays for a compensable cause, such as owner delays.

Change Orders and Proposed Change Orders

A written change order is usually a bilateral agreement adjusting the contract. If the change order provides for no extension of contract time and the contractor signs off, without reservation, for a fixed-dollar sum, normally the contractor cannot later make a delay claim for the work covered by the change order. The legal theories supporting this concept are release, settlement, and accord and satisfaction.

Requests for Information

Requests for information may help identify errors or omissions in the contract documents. An excessive amount of time to reply or resolve the request for information may form the basis for a delay claim by the contractor.

Job Cost Information (Unless the Claim Is for Liquidated Damages at a Set Rate)

Most construction companies use computerized job cost reporting systems. Often, the reports compare actual costs to budgeted costs by cost codes. If the contractor can show actual damages, the actual costs should be used. On the other hand, a defendant may be able to prove no or minimal damages actually were incurred by studying the job cost data.

Sometimes the best documentation concerning delays may be found in the files of the other side or third parties. When a party claims delay by my client, I especially like to look for documents where that party blames others for the delays. I have had cases representing the contractor where the owner's files include letters blaming its architect for slow responses to questions or design problems. I have also represented the owner where the contractor's files reveal delay notices to its subcontractors for the same delay period as the one being claimed against my client.[4] Finding the dirty laundry of the other side can sometimes help settle the case faster and, if not, provides a sound defense for my client.

Developing a Theory of the Case

Once I have gathered the facts and documentation, have reviewed the contract, and understand the applicable law, I develop the theory of the case. The theory of the case is a story about how my client is correct about the dispute. In a delay claim, if I represent the plaintiff, the theory explains the delays, why they are compensable, and how they resulted in damages to my client. The theory needs to be straightforward, consistent with the facts, provable with backup documentation, and capable of withstanding likely defenses. If the contract contains a "no damages for delay" clause, for example, I must develop the theory in a way so the claim can fit within an exception to the clause.

One such exception might include extended home office overhead costs. Historically, contractors have calculated these damages using the *Eichleay* formula.[5] In contracts with the federal government, however, recent case

---

[4] See Appendix I.

[5] The *Eichleay* formula grew out of a case at the Armed Services Board of Contract Appeals, *Eichleay Corp.*, ASBCA No. 5183, 60-2 BCA ¶ 2,688. The formula is as follows:

1.   <u>Total Corporate Overhead (G&A Expenses) during Total Contract Period</u>
      Total Corporate Billings (Income) during Total Contract Period
  x   Total Contract Billings = Total Overhead Allocable to Contract

2.   Allocable Overhead ÷ Total Days of Performance = Daily Overhead Rate

3.   Daily Overhead Rate x Days of Compensable Delays = Extended Home Office Overhead

law has substantially narrowed the circumstances in which the *Eichleay* formula can be successfully employed.[6] If I am assisting a contractor in preparing a delay claim on a federal project, I make sure the right facts are present before asserting an *Eichleay* formula claim. There are alternative methods I explore that may be more successful in the end. These include charging certain impact costs as direct costs, not as delay or overhead costs, and recovering home office overhead as a percentage markup on all direct costs.

As another example, a contractor trying to deal with project delays may add a late shift or put its crews on overtime hours. These efforts may save the project time but at an added cost. One might think of these as "acceleration costs." To recover for acceleration, however, the owner usually needs to issue an order to regain time or threaten some consequence for the project slippage. If this element is missing, I would explore whether the costs might still be recoverable as an effort to mitigate the impact of the original delay events for which the owner was responsible.

No one size fits all. Accordingly, I am prepared to tailor the theory of the particular case to the circumstances. I am also flexible enough to tweak the theory as new and different facts emerge.

Finally, I always keep in mind my audience when preparing a case. If the decision-maker is an arbitration panel of construction industry leaders, I present a more fact-intensive case. If the decision-maker is a jury, I keep in mind the need to educate them about complex construction issues. A more simple presentation may be more appropriate.

Troubleshooting: Avoiding a Problem in the First Place

Don't Wait to Read the Contract until There's Trouble

Some clients spend too little time at the time of contracting reading the contract and understanding its requirements and the allocation of risks.

---

[6] *P.J. Dick Inc. v. Principi*, 324 F.3d 134 (Fed. Cir. 2003). In *P.J. Dick*, the Federal Circuit held both that the *Eichleay* formula is the exclusive means to be used for compensating contractors for delay claims on federal projects and that there are several conditions to be met before a contractor is awarded *Eichleay* damages.

This is somewhat understandable when dealing with fixed bid projects where the contract is offered on a "take it or leave it" basis and the contract documents include reams of boilerplate clauses in small print. But a contractor who bids on a project of any substantial size needs to understand what he or she is getting into. A review and evaluation must be done of whether an unacceptable level of risk has been shifted to the contractor by the terms of the contract. These include but are not limited to such clauses as "waiver of consequential damages,"[7] "no damages for delay,"[8] extreme warranties, unreasonably short notice periods, indemnification of others for their own negligence, and the owner as the binding decision-maker for disputes.

If, on the other hand, the contract is offered on a negotiated basis, the unreasonable terms should be discussed and hopefully revised in a more balanced way. But regardless of the method of procurement, each party involved must read the contract and understand their respective obligations *before* the commitment is made.

Once the contract is entered into, and before performance begins, each side must be familiar and ready with the requirements concerning notice, decision-making, payment, claims, and disputes. Botched handling of these obligations inevitably compromises the strength of the claim or defense.

Document the Problem When It Arises

Participants in the construction process often bemoan the fact that they are there to build the building, not document a claim. Unfortunately, claims often are not resolved until long after the project is finished. The winner of the dispute is frequently the side with the better documentation. As an arbitrator, I give greater weight to contemporaneous project records than to unsupported witness testimony given months or years after the events in question. We have also polled juries and they agree that written evidence made at the time can be critical to their verdict. Finally, lack of timely written notice may violate a contract requirement and cause a judge, as a matter of law, to dismiss the claim.

---

[7] See Appendix J.

[8] See Appendix K.

Know the Contract's Limits

Sometimes owners are too risk-averse and shift too much of the risk of their own failings to the contractor. A good example of this overreaching is a "no damages for delay" clause that excuses the owner for its own delays and interferences.[9] My owner clients sometimes place too much faith in these exculpatory clauses. Courts and arbitrators narrowly construe the clauses. In fact, several judicially crafted exceptions to the clause have been recognized, such as delays due to active interference of the owner, delays beyond the contemplation of the parties, delays amounting to an abandonment of the work, and so forth.[10] In the right decision-maker's hands, the exception could eat up the rule and the owner, to his or her great surprise, is left exposed to the contractor's delay claim.

Most construction contracts require that all changes in the work are to be authorized by a written change order and that the contract may be modified only by a writing signed by the parties. Cases, however, have held that all contract clauses may be modified by the conduct of the parties or by oral agreement, even one that says no modification may be made absent a signed writing. A pattern of processing and paying for changes when no prior change order is issued may show a waiver of these contract provisions.[11] The lesson is that certain provisions of the contract may be enforced or not depending on factors such as the circumstances, the forum, and the equities of the situation.

---

[9] See Appendix K.

[10] See *Corinno Civetta Construction Corp. v. City of New York*, 67 N.Y. 2d 297, 502 N.Y.S. 2d 681, 493 N.E. 2d 905 (1986); *State Highway Admin. v. Greiner Eng. Sciences Inc.*, 83 Md. App. 621, 577 A.2d 363 (1990); see generally Kovars & Peters, "'No Damages for Delay'" Clauses," Edition II, *Construction Briefing Papers* (Federal Publications, 2000)

[11] See *Taylor v. University Nat'l Bank*, 263 Md. 59, 282 A.2d 91 (1971); *Freeman v. Stanbern Const. Co.*, 205 Md. 71, 106 A.2d 50 (1954); *Beatty v. Guggenheim Explor. Co.*, 225 N.Y. 380, 122 N.E. 378

Avoid Communicating to Explore Resolution before Positions Harden

Many construction law disputes arise out of interpersonal issues. When parties to the contract keep lines of communication open and each shows a willingness to listen to the other and to compromise, the likelihood is great that the problem will be worked out before it "goes legal." I advise my clients to explore resolving problems whenever possible. Sometimes it helps to have the principals on each side discuss a process, at the beginning of the project, for resolving disputes that are not settled at the field level. There are a number of alternative dispute resolution methods that can be employed, including mediation, binding arbitration, mini-trial, or dispute resolution boards. Sometimes simply having the principals meet can help work through the problem. If possible, get the alternative dispute resolution method agreed to in writing before you are faced with a dispute and angry project participants.

Conclusion: Keys to Success

The keys to my success are preparation, attention to detail, and communication. I always strive to be as prepared as possible before I talk to my client. That means reviewing my file, researching the issues as appropriate, coming to the meeting ready to ask questions, and showing the client I care about them and their problems. Attention to detail is also critical. It guides my determination of strategy and my preparation of pleadings, mediations, and trial work. My experience tells me that being more prepared than my adversary and understanding all the fine points of the dispute can translate into victory, whether in settlement discussions or trial. So-called "stupid questions" are asked without hesitation until I gain a full understanding of the matter. The final element, communication, is key to any successful relationship, especially the one between attorney and client. I try to inform my clients regularly about my strategy, my efforts on their behalf, and my evaluation of the case, and I copy my clients on virtually all communications with my opponent or the dispute forum.

Additionally, my success is largely impacted by my ability to keep up with the constantly evolving issues of the legal and construction industries. Therefore, I attend a number of meetings each year of the American Bar Association's Forum on the Construction Industry. The forum is the largest association of construction lawyers in the country.[12] Excellent educational programs are presented at each forum meeting. I am also an active member of the forum as a steering committee member of division nine. Finally, I speak and write regularly on construction law topics, which also helps me stay current with the issues. I also read approximately five periodicals on construction law, two on litigation, and two on the construction industry on a regular basis.

Joseph C. Kovars is co-chair of Ober, Kaler, Grimes & Shriver's construction practice group. An experienced civil litigator, he concentrates his practice in construction and public contracts law.

Mr. Kovars has served as first-chair litigator in large and complex cases, with appearances before federal and state courts, trial and appellate levels, boards of contract appeals, and the comptroller general. He has extensive experience with contract interpretation disputes, delay and impact claims, contract terminations, bid protests, mechanics' liens, bond claims, and injunctions. He has participated in a number of arbitrations both as an arbitrator and an attorney.

Mr. Kovars represents contractors, subcontractors, sureties, and public and private owners. He is currently active in several complex arbitrations, mediations, dispute boards, mini-trials, and other alternative dispute resolution proceedings as arbitrator and attorney representative. He is listed in The Best Lawyers in America *and has an AV rating from Martindale-Hubbell.*

Mr. Kovars earned his B.A., magna cum laude, from Boston University and his J.D., with honors, from George Washington University's National Law Center.

---

[12] The Web site of the American Bar Association's Forum on the Construction Industry is www.abanet.org/forums/construction/home.html.

Client Relationships and Strategies in Construction Disputes

Mark E. Hills

Partner

Varnum, Riddering, Schmidt & Howlett LLP

The Role of a Construction Lawyer

My work as a construction lawyer includes two major components. The first is prosecuting or defending litigation matters relating to construction projects and contracts that have become problematic. The second is advising clients on construction projects and contracts they are considering or into which they are entering.

The principal portion of my practice as a construction lawyer is in the litigation of problematic construction projects and contracts. In this regard, my focus is on establishing the client's compliance with the construction documents and working to maximize recovery or minimize exposure as efficiently and economically as possible.

Major Construction-Related Issues Clients Need Help With

The most significant construction-related issue with which clients need help is litigation of defective construction issues, which includes poor workmanship, violation of industry standards, use of substandard materials, and unauthorized departures from design documents and specifications. Secondly, clients often seek help with litigation relating to construction finance issues including unauthorized change orders, cost overruns, payment disputes, bond claims, and construction/mechanics' liens. Another type of litigation is that pertaining to design defects, which encompasses breach of professional standards, breach of industry standards, and cost-related issues. Finally, clients often need assistance with the formulation of construction contracts, such as negotiations relating to technical performance issues and procedural matters.

Construction Law's Various Components

The different components of construction law are contract negotiation and formulation, design, contract execution, and dispute resolution. My firm has the ability to assist clients in a wide variety of financing alternatives, whether through a typical mortgage financing for private projects or bond issuance for public projects.

We are also able to provide services with respect to each component of construction law. First, contract negotiation and formulation is critical to construction law, as it provides the foundation for the entire project. With respect to this component of the law, I work to establish the most advantageous contract for a client both in terms of its execution as well as enforcement in the event disputes arise. For instance, the local construction industry typically uses the standard contracts and general conditions documents provided by the American Institute of Architects or the Engineers Joint Contract Documents Committee. However, the various clauses of those documents do not apply to all situations. After a bad arbitration experience, a client may want to strike the arbitration and mediation provisions from a general conditions document. Similarly, a contractor may be unwilling to proceed with a project if the owner insists that the waiver of consequential damages clause be stricken. Given the wide range of topics addressed in these documents and the substantial impact they may have, careful review of contracts and discussion of hypothetical situations is mandatory in the negotiation and formulation stage.

Aside from contract formation, I am not actively involved in the design phase of construction projects, nor am I typically involved in the contract execution phase except as disputes may arise. However, dispute resolution through negotiation, mediation, arbitration, or court litigation is the central aspect of my practice. With respect to this component, I work to establish the supremacy of my client's position while searching, most importantly, to determine the truth of the dispute. This requires a detailed understanding of both large-scale issues and items of more limited detail, and it often entails significant time reviewing the two ultimate sources of evidence: testimony from individuals (gathered initially through interviewing witnesses) and information contained in documents. A thorough understanding of issues and evidence is necessary to begin establishing a case strategy as well as strategic plans for addressing areas of potential difficulty. This also provides the critical background needed for viewing a dispute in an objective fashion, thereby enabling counsel to determine whether his or her client's position is one that may prevail.

Financial Implications of Construction Law: Adding Value for Clients

A common effect of most construction disputes is cost escalation, whether it appears in the context of actual construction cost overruns or costs associated with the various dispute resolution mechanisms. Accordingly, careful negotiation and formulation of contracts and early work to resolve disputes prior to degradation of business relationships provide a critical benefit to clients: cost avoidance.

Where dispute resolution is necessary, efficient and zealous prosecution of a client's position often leads to either quick resolution or significant recoveries, thereby leading to benefit maximization.

I proactively add value for clients by consistently updating them on changes in the law that impact the manner in which they do business. For instance, Michigan is one of the minority of states that recognize the validity of "pay when paid" clauses in construction contracts. Such clauses effectively bar a subcontractor's ability to collect its fee from a general contractor until the general contractor has received payment from the owner. We have been able to avert potential litigation or achieve rapid success in litigation that is already commenced by ensuring that general contractor clients include this clause in each of their contracts. From a retroactive perspective, we add value to our clients by utilizing the results from disputed matters as object lessons. If, for example, particular aspects of a client's billing practice are the subject of dispute, we continue to work with the client after the matter has concluded to suggest appropriate modifications to its practices in order to avoid similar problems in the future.

Common Client Mistakes with Respect to Construction Law

An area that consistently surfaces as a problem for clients is the process of managing a construction project or contract from a financial perspective. Historically, the construction industry is one in which a project went forward on nothing more than a handshake. That has changed to some degree, but projects often become problematic when the parties to a contract fail to follow prescribed procedures for approval of changes to the contract through written change orders or directives on the belief that everything will "wash out in the end." What typically washes out is the formerly optimistic relationship between owner and contractor as well as any short-term economic benefit from the project as a result of significant legal expenses incurred to resolve the disputes. These disputes could be easily avoided if an insignificant amount of additional time and expense were expended in properly documenting changes to the parties' contractual obligations.

Achieving Success with Respect to Clients

The most critical component of achieving success with clients is the ability to act proactively with clients to avoid problems altogether. Alternatively, a successful construction lawyer should be able to devise mechanisms to avoid repeating mistakes. Another important element of achieving success is managing client expectations when problematic matters cannot be resolved and litigation ensues.

With respect to achieving success in this industry, it is also important to stay on top of one's knowledge of current and emerging trends, cases, and laws, which I do by participating in the construction industry generally. My firm is an associate member of Associated Builders and Contractors, a leading trade association. Regular attendance at its meetings and conferences and review of trade publications help keep me up to date on issues from within the industry. Similarly, participation in seminars as an attendee or presenter assists in expanding our knowledge base of construction issues. Beyond the scope of the construction industry, we constantly monitor judicial decisions and legislative changes that impact the wide variety of issues that exist in the construction industry. We also participate in legal industry groups such as the American Bar Association's Forum on the Construction Industry, which provide seminars and written materials addressing legal issues within the field of construction.

Preparing for the First Client Meeting

The first step I take before meeting with a client is to ensure that there is no conflict of interest in the potential representation. Once this is established, the nature of preliminary information and research depends on the scope of my proposed representation of the client. Typically, I focus my attention prior to a first meeting on the identities of the involved parties and obtaining, indirectly, as much objective information about them as I can. This information gives me the ability to set my own rough expectations regarding the nature of the parties' playing field before I acquire more subjective and direct information. It also gives me the opportunity to determine if any of the parties have established practices or reputations that may impact the substantive issues at hand.

The First Client Meeting: Information to Obtain

Upon meeting with the client, I verbally gather as much information as possible, inquiring as to: (1) any historical relationship between the parties to the dispute; (2) the history of the project in dispute; (3) the involvement of other parties, if any; (4) the specific aspects of the project that are in dispute; (5) the aspects that are not in dispute; (6) the client's understanding of its opponent's position; (7) the client's statement of its own position; (8) any problems with its position of which the client is aware; (9) pertinent individuals with knowledge of the disputed events or issues; and (10) documents containing information relative to the dispute. This gives me the client's overall perspective on the dispute, highlights potential points of bias, identifies issues that are critical to the client, and begins to establish the client's expectations. Further, it gives me the ability to continue to identify individuals who possess information relating to the dispute.

I then work to gather as much supporting information as possible, typically by obtaining contracts, blueprints, construction specifications, correspondence, e-mail, and other documents relating to the project and specific dispute. Both areas of information are critical to the initial assessment of the dispute and formulating an initial strategy.

The Process of Formulating a Client Strategy

Before formulating a strategy, one point must be made clear to a client. It is my job, as counsel, to provide alternatives and information about those alternatives. The client must then make the ultimate decision as to the course of action to pursue and, as often occurs, when to modify its position or strategy. It is inappropriate for counsel to interject himself or herself into the decision-making process once counsel has provided the various options or alternatives and all pertinent information associated with them.

Appropriate assessment of a client's goals is critical to determining a client's strategy. Factors that should be considered include: the nature of the harm the client potentially faces, whether the resolution is time-critical, the nature of the client's overall financial situation, the magnitude of the potential loss or recovery, whether the dispute involves matters of principal or particular symbolic significance, and the overall impact of the dispute. If it appears a client is pursuing goals that are imprudent or unethical, the representation will be declined or, if the representation has already commenced, counsel should terminate it.

The nature of the harm the client potentially faces is critical. If a client is threatened with a temporary restraining order that could shut down an entire project, the potential damages are extraordinary. Therefore, a client must decide almost immediately whether to work from a negotiation or a confrontational perspective.

This same choice must be made in instances where a project is at a time-critical stage. If, for example, the construction project is complete and payment is withheld, it may be less critical to the project owner to take an aggressive position. Alternatively, under the same scenario, the contractor may be suffering significant financial distress as a result of withheld payment and may need to push its position strongly. This is especially true if the contractor is in a position where withheld payment may interfere with its viability and ongoing business operations.

The magnitude of the potential loss or recovery is closely tied to the foregoing issues as well. If a claim by a licensing agency creates the possibility of a long-term loss of business, a client may have no choice but to zealously challenge the agency. If, on the other hand, a fifth-tier material supplier with a claim of less than $5,000 has appeared, it likely makes sense for the client to promptly negotiate the best resolution possible without incurring significant legal expenses.

Matters of principle are often the most costly from a legal perspective, because clients do not apply business analysis. As a result, it is important to regularly discuss the matter and encourage evaluation from a business perspective rather than from a position driven by emotion or symbolism. Ultimately, the big picture must be kept in mind. Taking advantage of a subcontractor may provide some short-term economic benefit, but once the word of such action is out in the local market, the long-term detriments can be devastating, as word of this conduct will most certainly spread within the local community and create the risk of lost future business or business relationships. Decisions, therefore, should not be made in a vacuum. It is important for counsel to provide advice accordingly.

Finally, it is important to note that the side my client is on in a construction dispute does not impact the way I decide to proceed. The general approach remains consistent regardless of whether I am working with a plaintiff or a defendant.

Building Positive Client Relationships

I believe the most critical aspect of a positive client relationship is honesty. An attorney for whom I worked some years ago used to say to clients, "There are two people you don't lie to: your lawyer and your minister. And I don't give a crap what your minister thinks." Until there is absolute honesty—a true two-way street—there cannot be genuine trust, the second cornerstone to a positive client relationship.

I expect my clients to tell me everything, good and bad, about their dispute. I will not tolerate, nor will I continue to represent, a client who withholds information from me. Doing so only guarantees ultimate dissatisfaction or worse. At the same time, I give my clients my honest opinion, good and bad, about their dispute and the approach I think makes the most sense to achieve the strategic result they desire.

Handling "bad facts" varies greatly from one disputed matter to another, and the manner in which they are handled can change depending on the aspect of the dispute to which they relate. Once bad facts are identified, they can be addressed through methods such as working to establish that actions or inactions of the opposing party actually caused the factual occurrence, or refocusing the strategic approach to legal theories in which the bad facts have little or no significance.

Ultimately, a client's attitude has little impact on the voracity with which I proceed in any given case. I have heard the "I need a pit bull" statement from a number of clients. This is presumably based on too much television and the mistaken belief that a loud, obnoxious attorney will achieve the best result. Those results, in fact, are best achieved through thoughtful consideration of the facts and alternatives, a sound strategic approach to gathering information and facts, and skillful presentation of the client's case at the appropriate time, none of which require a pit bull approach from an external consideration.

I recall one matter specifically where the client did not believe we were acting aggressively enough but was willing to listen to continuing counsel regarding the strategy we proposed. We ended a two-week arbitration hearing with a $2.1 million recovery representing 97 percent of the client's total damage claim. While the same result would likely have been achieved through a more aggressive posture, it would have been obtained at higher legal cost to the client, thereby reducing its actual recovery.

The Impact of a Client's Willingness to Settle on Strategy

The possibility or probability of settlement should affect strategy only if it is the strategic goal from the start. Settlement, however, is a two-way street. Since settlements often do not occur until one of the dispute mechanisms commences (and, ordinarily, later in that process), a strategy must be developed to move the matter forward to completion with settlement as one possible alternative.

Clients often take the early position of refusing to settle. However, this attitude ignores the wide range of processes now in place in the judicial system that encourage parties to settle, often under the threat of the possibility of paying some or all of the opponent's legal costs.

Regardless of a client's early approach, the settlement mechanisms do work, and clients in construction disputes typically do settle. First, as I indicated previously, most of the dispute resolution mechanisms are lengthy. We normally advise clients that, in the local court system, their first trial date will be scheduled fourteen to eighteen months after a suit is filed and that cases typically do not go to trial on the first trial date. Arbitration hearings usually occur approximately one year after commencement of the proceeding. This significant commitment of time, energy, and resources over long periods often wears down even the most zealous client who would understandably prefer to commit themselves to the business of making money.

Additionally, and partially as a result of the procedural mechanisms governing the dispute resolution process, litigation is very expensive. The client who was originally so opposed to settling may have a different perspective after going through the process and understanding the costs involved.

Presenting Options to a Client

Litigation and the other dispute resolution mechanisms are fluid. They involve regular re-evaluation of the client's position and modification of both litigation and settlement strategies. Therefore, it is important to remain open to a multitude of alternatives and to lawyer creatively when necessary. If a client is struggling financially and desires to eliminate the monetary risk posed by an ongoing dispute, it may be most advantageous to insist on early attempts to reach a settlement rather than taking a highly confrontational approach. Alternatively, a financially sound client may have no significant motivation to resolve a dispute early in the process.

Though providing alternatives and assessment of alternatives should be the limit of counsel's function, those discussions are often not concrete enough for a client to serve as the basis for a decision. Clients are sometimes looking for more. In these instances, I have found that anecdotal information can provide that additional information. While the facts of each construction dispute obviously vary, there are few disputes with which counsel are not familiar. Associated anecdotal information can provide a more concrete example of the positive or negative ramification of a certain strategy and assist in the decision-making process for the client. When the client is ambivalent to his or her own construction dispute, these anecdotes personalize the situation and often shift the client's perspective and mood to a more decisive one that is far more amenable to progressing toward success with respect to the case.

The Biggest Mistakes Attorneys Make

I think a significant mistake attorneys make is failing to regularly look at a matter as objectively as possible. I am hired to be an advocate and to advance my client's position zealously. That does not mean, however, that I lose sight of the proverbial forest while staring at a tree. If I am unable to take off my advocate's hat and look at the matter with relative objectivity, I will not be able to devise an appropriate strategy and then, as necessary, modify that strategy as a dispute proceeds. The client is the one most likely to be caught up in emotion and indecision. It is my role to provide advice subjectively as well as objectively.

In this regard, one benefit of working in a larger firm is the ability to find an uninvolved attorney down the hall and solicit his or her general thoughts as a means of checks and balances. Reaching out in this fashion helps avoid tunnel vision on a particular matter and encourages more objective thinking. This has been an effective strategy when addressing the challenges of keeping a fresh and objective perspective on a case in which I have been immersed, and I think all construction lawyers should strive to achieve some level of detachment and rationality through a similar process during each case.

Mark E. Hills has been a partner with Varnum, Riddering, Schmidt & Howlett LLP since 1998. His litigation practice areas include construction, real estate, condemnation, and commercial matters. Previously, he was an associate with Farhat, Tyler & Associates PC from 1994 to 1998. Prior to that, he was a law clerk for the Honorable Lawrence M. Glazer of the Thirtieth Judicial Circuit of Michigan.

Mr. Hills earned his B.A. in political science in 1989 from the University of Michigan and his J.D. in 1992 from the University of Toledo College of Law. He is a member of the American Bar Association, the State Bar of Michigan, the State Bar of Wisconsin, the Grand Rapids Bar Association, the Ottawa County Bar Association, and the Associated Builders and Contractors.

Establishing and Achieving Objectives

John W. DiNicola II

Partner

Holland and Knight

I have represented just about every entity that touches a construction project during my years of practice in construction law, although I represent general contractors more often than subcontractors, sureties, or owners. My firm assists clients throughout the entire life of a project. We draft and negotiate contracts. We also assist our clients during the project whenever issues arise such as claims, delays, or other issues. "Project counseling" happens quite often, especially if a client is involved in a troubled project, one with many construction or contract issues and conflicts. After the project is completed, we assist our clients with dispute resolution involving owners, subcontractors, insurance companies, and others.

Construction law attorneys offer the greatest value to their clients by analyzing the big picture that surrounds any construction project and by offering valuable advice so the client can make the best decisions possible. Often, the client is only focused on today and the issue that is on his or her desk at that particular time. It is often difficult for the client to take a step back to review the overall project status, including its liabilities, and to take another step further back to analyze how this project is affecting the company as a whole, financially and in a business sense. A skilled construction lawyer can develop that big-picture analysis to try to guide the client through issues and toward a good result.

Contract Formation and Dispute Relation

On any construction project, there are dozens of contracts such as owner and general contractor contracts, owner and architect contracts, subcontracts, bonds, and purchase orders. One of the most common construction-related issues requiring the assistance of an attorney is contract formation and negotiation. If contracts are not set up properly with risks allocated appropriately among the owner, the general contractor, and subcontractors, the likelihood of disputes arising at the end of the project increases. It is important for the success of the project for all parties, and for the general contractor in particular (because the general contractor enters into more contracts than any other entity for a construction project), that all of the contracts are consistent in how they allocate risk, address changes in the field, and address payment, for example.

However, a major focus of our practice involves a full range of dispute resolution techniques—claim development and negotiation, mediation, arbitration, and litigation. Our involvement in dispute resolution often starts with project counseling on changes or scheduling delays that occur during the project. We assist clients in drafting claims and compiling proper documentation and then attempt to resolve those claims with the owner. If the claim cannot be resolved through negotiation, we then assist the client in resolving the claim through mediation, arbitration, or litigation.

Mediation is a very popular form of dispute resolution, wherein the disputing parties engage a neutral party, called a mediator, to assist the parties in negotiating a settlement of a dispute. While mediation is now very widely used in all forms of litigation, the construction industry was a pioneer in this dispute resolution technique. As construction attorneys, we represent our client in mediation, presenting arguments and responding to issues as they arise both from the opposing party and from the mediator. Mediation is a great tool to employ primarily because the parties are in control of their own fate—the dispute will not be resolved unless all parties agree. In contrast, in litigation or arbitration, a judge, jury, or arbitrator will determine the outcome.

Common Mistakes

Poor contract documents are a common mistake in this industry. Claim and dispute avoidance is an important aspect of the construction attorney's role. The first thing that should be done when a construction project starts—but often is not—is ensuring that the contracts are fair, distribute risk equitably, and work well together. If improper contract formation leads to differing obligations among the parties, the project is set up for disputes down the road. The second biggest mistake that is made in construction projects is failure to follow contract procedures. Contracts in construction are very detailed. They have been developed over several decades to deal with certain situations that frequently arise during a construction project, such as differing site conditions, hazardous materials, and claims for delay. Specific clauses in the standard forms address each particular issue, but each standard form is subject to negotiation. A contractor must comply with contractual time limits and other substantive requirements in order to resolve disputes or at least preserve the right to assert claims for extra work, changed conditions, and schedule delays, for example. However, in many cases the people who are working on the project are unaware of the time limits and procedural requirements for asserting claims. Because of that disconnect, a client can have a claim rejected for failure to follow procedural requirements, even though substantively the client should be entitled to the claim. It is, therefore, important for all project personnel, including field personnel, to understand the notice requirements of the contract.

A third mistake that often occurs in troubled projects involves a failure of upper management to oversee the project. Often, upper management is too busy to notice potentially troubled projects that may be leading toward disputes. In some cases, upper management does not know about the situation until the disputes get to a point where lawyers have to get involved. If upper management can institute proper controls or monitoring procedures, they can often sense potentially troublesome issues before those issues become claims that are expensive and time-consuming to resolve. To control this issue, it is helpful for upper management to encourage frequent and honest communication with project management staff.

The final issue that most often poses trouble for clients involves strategic planning. Many of our clients in the construction industry, especially contractors, focus on what their cash flow and financials are today without taking a step back and looking at what may happen one month, six months, and one to two years down the road. It is helpful for the client to implement a strategic business plan and consider its financial implications. Most successful contractors understand their markets, understand their strengths, and remain consistent between their business strategy and their focus markets and strengths.

Successful Strategies

Two important strategies are essential to successful client representation in this field. First, the attorney needs to take a big-picture approach. A client may call because of a problem with a subcontractor on a particular project, focused on that particular subcontractor and that particular issue. While it is important to address that issue, the attorney also needs to analyze the project as a whole. What other issues are out there? What is the relationship with the owner and architect? What are the project's financial and construction status? Once you have understood that information, you can make better decisions and provide better advice with respect to the issue that is on your desk that day. My second key strategy is to listen to my client—to what the client perceives the issues to be—and then try to elicit from the client what its goal is in terms of resolving this particular dispute. This is often difficult to do because the client may not really know his or her goals or may not know how to articulate them. It is important for the attorney to ask appropriate questions of the client to develop and understand the facts so the attorney and client better understand the issues. Once you have a clear idea of the client's goal, you need to develop a strategic plan to achieve that goal that addresses both the current and long-term issues facing the project.

Setting a successful strategy also involves staying on top of your knowledge in this field. There are several periodicals that come out on a weekly or monthly basis that I read regularly. The American Bar Association has a separate subgroup for construction lawyers called the Forum on the Construction Industry that publishes outstanding materials I review on a regular basis. These materials and other attorney groups focus primarily on legal issues affecting the construction industry. I also attend the forum's biannual meetings as well as other groups' meetings. The construction industry itself also publishes periodicals that are geared not toward lawyers but toward construction entities—general contractors, owners, architects, engineers, and subcontractors. I read those as well to keep up to speed on what is new and what the economy is doing.

Preliminary Research

When I first start working on a construction law project, I call the client and have a short discussion. I try to gauge the overall status of the project in order to determine whether the issue is something that needs immediate attention. In other words, is this issue driving the project, or are there any bigger problems we need to address? In conjunction with these answers, I determine the client's objective and decide whether it is feasible.

Getting the information you need can be difficult, depending on the client. By and large, people in the construction industry will tell you exactly what they want. They are very direct. The real issue is getting all of the information that will allow you to provide the best advice possible—that is the challenge facing the construction lawyer. As in any other industry, your client will often tell you what reflects well on them, but they will not tell you, at least initially, what is bad. Only when you have all the facts can you help them make informed decisions.

During the initial contacts, I need to learn exactly what type of work the client does. The construction industry, like the legal industry, is getting more and more specialized. I like to know whether the client is involved in vertical construction as opposed to heavy or highway construction. Do they do private or public work, or a combination of both?

I also try to find out about their litigation history. Some company cultures are more aggressive than others. If the client has a history of settling disputes and avoiding adversarial situations, you need to work with that client's personality and culture while still providing the soundest advice you can. If the client is one that stands on principle often, the attorney likewise needs to understand that culture. Although the advice would not materially differ with either culture, understanding that culture permits the attorney to choose the best way to approach the client so the issues and options are discussed and understood.

Formulating an Effective Client Strategy

The first step I take in terms of formulating a client strategy is aimed at really understanding the immediate issue, which can take some time. It often requires persuasive skills to draw information out of the client. It is important to ask the right questions in the right way to get the information and documents. The attorney also needs to use his or her experience to ask for the most appropriate information because, for example, it may not be a good use of your time or the client's resources to look at all the project records when the client has an issue with just one subcontractor.

Next, you must take a broader view of the issue. Look at the project as a whole and learn your client's ultimate goal—not just with this issue, but with the overall project. When determining the client's actual goal, you can adjust that goal to meet realistic expectations. When we are engaged in litigation, which is an adversarial forum, every lawyer zealously advocates for his or her client, but there is a separate facet of the construction lawyer's role, which is to provide counseling and an objective analysis of a particular issue and how it affects the overall project. Only then can you answer the question: What issues do we need to deal with now, and how are we going to deal with them as the project goes on?

When formulating client strategy, it is important to keep in mind that there are a lot of political considerations connected with every dispute. For example, what is the client's relationship with the owner? If your client is a general contractor and the owner is a private entity, has the client already performed several projects with the owner, and does the client anticipate performing future projects with that owner? If that is indeed the case, you may take a different view with respect to disputes with the owner than you would if this was, for example, a public job where future projects are subject to public bidding laws.

You also have to look at the client's relationships with the architect and the subcontractors. Your client lives and works with these entities every day, and you do not. As a lawyer, you may be working on only one project at this point in time for that client, and you may not understand all of the implications of the decisions that will be made with respect to this particular dispute. Therefore, it is very important to understand your client's business and its business relationships.

Dealing with "Bad Facts"

There are two sides to every coin. Once you start gathering information, you are going to assimilate facts that are both good and bad for your client's case. You have to tell the client which facts are good, which are bad, and which can be interpreted either way. This objective analysis allows the client to understand the issue from a business perspective. It is important to keep the client informed every step of the way as the facts are developed so he or she can make the right business decision at any particular time during the dispute resolution process.

Every case has weaknesses. Generally speaking, the attorney will point out that the weaknesses are not as weak as they may seem initially or that they are not relevant to the dispute. I always find it is better to acknowledge the weaknesses in your case rather than ignore them, because if you ignore them it only makes the opposing party's case stronger.

Determining Project Status and Other Assessments

It is always important to determine the project status (i.e., completion status) early on when undertaking a construction law dispute. If we eventually decide to enter a dispute resolution procedure, one of the non-quantifiable costs of litigation/arbitration/mediation is that it takes manpower from the client to participate in the process. If a project manager has to spend eight hours with the attorney, he or she will not be able to spend that time working on a project. Therefore, if the client is in the middle of a project, you have to consider whether it is worth dedicating the client's personnel to the legal process at that particular time at the expense of the project.

The non-legal ramifications of a client's construction matter are very important for the lawyer to understand and consider when counseling the client. Our clients are not in the business of litigation. Although the client wants claims and litigation resolved favorably, it is not part of their overall strategy. Once lawyers get involved in a contentious dispute, it is going to consume a lot of the client's time in terms of getting information from project managers and superintendents—time that could be spent on other projects.

Another important consideration when analyzing the overall project is the relationship between the key players. A client may have a great claim with respect to an owner, but asserting that claim may jeopardize future business. The client may not want to risk millions of dollars of business two or three years down the road for a contentious dispute of far less dollar value.

Finally, you must address early document preparation of claims. I try to get my clients to think about how the claim is going to be presented to a third party. If the client can organize its field staff and project management to comply with the contract's notice requirements and properly document the project as it progresses, the client is much better off than if the client and attorney try to piece things together at the end. As a general principle, the party with the better and most comprehensive paper trail will have an advantage with respect to disputes later on, because memories tend to fade over time but the project records remain.

Financial Considerations

I also like to know the financial status of the project. Is the owner paying on time? Are there a lot of claims that have been rejected? What is the status of the subcontract balances? The answers to these questions assist in formulating an informed, objective analysis of any particular claim.

The client's financial status has a significant impact on case strategy. You need to know the financial status of a particular project, and the company as a whole, in order to develop the appropriate overall strategy for resolving a dispute.

It is important to consider the client's overall financial status rather than taking a project-to-project view to properly advise the client. What are the projected revenues? Is the client projecting to have cash flow during the year sufficient to absorb some issues that may arise down the road? Do any other financial issues exist? Understanding these issues assists the attorney in developing appropriate strategies and providing proper counseling in the short term and long term for the client.

Know Your Opponent

It is also essential to try to find out as much information as you can about the opposing party. If, for example, I am representing a general contractor who has a contract with an owner, it is important to know whether this is an institutional owner, a big corporation, or a developer. If it is a developer, it is usually a single-purpose entity and the only asset it owns is the building and land. There is a big difference in your strategic thinking if your contract is with a single-purpose entity that only owns one asset as opposed to entering into a contract with a corporation that has a lot of assets. Each case has three equally important aspects: entitlement, damages, and collectability. If your opponent only has one asset, it is essential to gain security in that asset prior to resolution of the dispute, which can usually be accomplished through mechanics' liens.

The Client/Attorney Relationship

Establishing a positive client/attorney relationship is an integral part of achieving success in construction law. It is important to be personable, understanding, and respectful with the client, and to not be condescending or demanding. The lawyer has to listen to what the client has to say.

Clients often become emotionally involved in disputes, and in such circumstances it is important to let the client vent, release the emotions, and then focus on the facts. When you listen, you can often get the information you need to successfully represent the client. A dispute or claim issue can be a distressing situation. The client is calling you because they have trouble and need help. The attorney must understand that, for most clients, this is not an enjoyable or familiar situation. You need to provide assurances and understand the emotions being expended by the client.

At the same time, a successful client/attorney relationship is based on honesty. You have to manage the expectations of your clients, which entails providing honest advice and objective counseling. To that end, the lawyer must maintain an independent, objective mindset with respect to dealing with the client, and keep the relationship within the realm of attorney/client privilege.

The client's attitude does not affect what I do substantively, because I have a professional responsibility to that client to do my work to the best of my abilities. If a client is very involved and helpful in terms of claim preparation, litigation, or arbitration, that helps me do my job better. If the client is not helpful, does not get me the information I need, and does not seem interested, that does disadvantage me to some extent. Nevertheless, lawyers have a standard to live up to, and no matter what the client says or does, we have to meet that standard.

It is disappointing from a personal standpoint if the client is ambivalent to its own construction disputes, but it is understandable. Clients want to build buildings. They do not want to litigate. Sometimes they may not feel the dispute is important to them on a particular day. Consequently, they may not provide you with the attention you believe the dispute deserves. It is our job as lawyers to understand their mindset, yet provide advice so our clients can make the best decisions they can possibly make regarding each dispute.

When providing advice, you must approach the client carefully. Your message is only going to be received if it is delivered in a way the client understands. I try to elicit what the client's concerns are with respect to this dispute and understand the overall situation the client is dealing with that day. It is also important to establish deadlines that are reasonable for the client and for me with respect to document production, information exchange, or other issues.

If the dispute evolves into an adversarial setting such as mediation, arbitration, or litigation, the attorney should obviously advocate for his or her client. When the attorney and client are discussing the case privately, it is important for the client that the attorney provides the client with objective analysis and advice as to the strengths and weaknesses in the case. Your case will be stronger if you have always been honest and objective when dealing with your client.

The Importance of Alternatives: Timing Is Everything

I always provide settlement options to my client. Nevertheless, the right to settle is the client's and the client's alone. Part of my job as a counselor and an attorney is to provide options to the client in order to resolve the dispute through either settlement or the adversarial process. If a client does not want to settle, I have to give that client my honest opinion as to the dispute so the client knows exactly what the risks are, including the costs of going forward.

There are many avenues to take with respect to dispute resolution, including settlement discussions, litigation, arbitration, and mediation—and timing is crucial. I always lay all of the options out on the table for the client. I then provide advice as to when I think it is the best time to implement any one of these options. For example, mediation is a very effective means of settling disputes, but the decision as to when to participate in the mediation, much like the timeframe for litigation or arbitration, is critical in determining whether mediation is going to be successful. Mediation is most effective when both parties understand the dispute fully and are committed to the process. Therefore, it is important not to engage in mediation too early when sufficient information exchange has not yet occurred, or too late when the parties have become too entrenched in their positions.

If the client wants to follow a strategy I do not think is in the client's best interest, I will tell them that. You have to be honest with the client and tell them exactly what you think based on your experience, and ensure that the client understands your concerns. It is then the client's choice as to how to proceed. If the client's choice is unethical, I simply will not do it.

There are going to be times when the attorney and client disagree regarding overall case strategy, but as long as they discuss the issues and the client understands the concerns of the attorney, the case can move forward.

Setting a Winning Strategy

The most important aspect of developing and implementing the right strategy in construction litigation is setting the objective. What does the client want to achieve at the end of the day? You cannot develop a winning strategy without having an endgame in mind.

The negative consequence of developing an improper strategy is putting the client in a worse position in terms of the dispute, the litigation, or its business than it would have been had the proper strategy been invoked. For example, if a general contractor has a dispute with an owner and a subcontractor, and you only consider the dispute with the subcontractor, you could make certain admissions or decisions with respect to that subcontractor that could jeopardize your position with the owner.

The only way to avoid such a situation is by getting all the information you need from your client—financial information, claim information for the overall project, and information regarding the relationships between the general contractor, the owner, and the subcontractor—and discussing the issues with the client. By doing so, you reduce the likelihood of taking a position in one claim that will jeopardize your position in another claim, putting your client in a worse financial position at the end of the day.

Armed with the right information, you can advise the client so the client can make an informed business decision as to how each individual claim will be addressed. If you make those decisions in a vacuum and without the proper overall context, the client can make decisions that may place it in compromising positions later on. Therefore, I always try to elicit information about the overall project claim status and then give advice that will limit the client's exposure as much as possible.

The attorney should maintain a big-picture approach with a project and understand that, in most circumstances, money is often the primary factor in dispute resolution. It is therefore important to determine where the money is and how to protect your interest in that area. You can then deal with any issues that may have lower significance. Simply put, if you just take on a project one issue at a time without understanding the big picture, you can jeopardize your position on other issues.

Achieving Your Goal

It's worth repeating: the biggest mistake you can make in construction litigation is failing to establish an objective. Without knowing what you want to have happen at the end of the day, you cannot develop a strategy. I often see cases where parties seem to be litigating just to litigate, and it is difficult to determine what they want the end result to be. It is vital to establish a reasonable objective and then develop a reasonable strategy to get there.

At the same time, a strategy hardly ever works out perfectly, so you have to be able to adjust it along the way. New facts will be discovered or there may be new circumstances outside the facts of your case that may affect how you approach a particular dispute (i.e., other issues may arise with the client or the opposing party may be having serious financial difficulties). Therefore, you need to set your objectives, make them reasonable, develop a strategy to get to your goal, but have the flexibility to adjust that strategy and maybe even the objective as you assimilate more information.

John W. DiNicola II practices in the area of construction litigation, arbitration, and mediation, as well as commercial litigation. He has extensive experience in counseling and representing owners, general contractors, subcontractors, and sureties on a variety of issues in the construction industry. In addition, he has significant experience in several forms of alternative dispute resolution, including arbitration and mediation.

Mr. DiNicola has lectured and published on several topics concerning construction law and dispute resolution for the Associated General Contractors of Massachusetts Inc., Construction Industries of Massachusetts Inc., the National Business Institute Inc., Lorman Education Services, and the Professional Education Systems Institute LLC. He has also lectured on construction law topics for Suffolk University Law School.

How to Approach Construction Litigation

Jonathan D. Herbst, Esq.

Partner

Margolis Edelstein

Construction law involves contract formation and interpretation, surety and indemnity law, claims procedures in the public and private sectors, and duties created by tort law.

Each stage of a construction project requires attention to legal implications. The design phase involves the owner, the design professional team of architects and engineers, and possibly construction managers. The determination of the method of project delivery and its financing will impact the relationships of the parties and their respective contracts. The contents of the general conditions, supplemental or special conditions, and specifications define the project and are an integral part of each party's contract.

The bidding phase on public projects requires adherence to federal and/or state law. The construction phase involves notices of delay, notices of changed conditions, notices of potential claims, the change order process, shop drawings, approved changes, payment, default, and dispute resolution.

To Settle or Litigate

As a litigator, my involvement in a construction law dispute begins only after a lawsuit or arbitration proceeding has been filed. An early settlement is usualy the best way to help a client involved in a construction dispute. A "win" after years of costly litigation is often insufficient to cover the actual costs to the client. In those cases in which a settlement is not possible, a good result often means a "loss" for a lesser amount than for what a settlement would have cost. More often than not, litigation is not a satisfactory method of resolving construction disputes. Trial usually results when one or more of the parties have overestimated their chances of success. To put it another way, the case has been improperly evaluated and the chance of obtaining more money or paying less money appears to be more likely obtained through trial than by agreeing to a settlement. It makes sense to go to trial only when your evaluation of obtaining a better result is estimated to be in the 75 to 80 percent range. The word "chance" is a very significant component of your evaluation. In this context, it means gamble. As strong as you believe your case may be, there is always the risk that the jury, judge, or arbitrator will not see it your way.

Avoiding Trouble

My father had a saying, "Never make the same mistake twice." Since my involvement in a construction project dispute begins after significant problems have developed, I always try to educate the client so repeat mistakes can be avoided.

The most common mistake I see made during construction law litigation concerns the acceptance of contracts. Subcontractors, in particular, tend to accept whatever contract is offered by the general contractor. Sometimes this is little more than a purchase order or an invoice with a "scope of work" or "proposal" attached. Another common shortcoming is the inappropriate use of standard contracts and forms. One size does not fill all. Frequently, general conditions are incorporated by reference that are inconsistent with or even contradictory to the terms of the original contract document. Reference is sometimes made to a standard form that is no longer in use and cannot be located. At other times, one set of standard contract forms is incorporated into a standard form adopted for use with an entirely different type of project delivery. In addition, unsigned contracts by one or both parties are frequently found in files with a notation that it is to be signed and returned.

There are two major causes of poor contract formation. First of all, contract details are not a priority for busy contractors. The bidding phase of a project is often hectic with predetermined deadlines that must be met. Contractors frequently submit bids on several projects with deadlines within days of each other, expecting that only a small percentage of bids will be accepted. When a contractor's bid is accepted, the pace becomes more hectic. The estimator is working on a different bid or bids, and estimating usually does not include contract formation. No project manager has yet been selected, because there was no certainty that the submitted bid would be the winning bid. A project manager is hurriedly found, often borrowed from another project until a permanent project manager can be assigned to the new project. Frequently, there is general knowlege within the organization that a new project has been won, but the company's resources have not yet been devoted to the new project. Deadlines, however, must be met. A proposed form contract is received and placed in the project file with no review or follow-up. The borrowed project manager returns to his or her original project, and his or her replacement begins solving construction-related issues assuming their predecessor has properly overseen the contracting phase. Only when a problem develops is it realized that the contract documents are not in order.

The contract is the "bargain." The contractor agrees to perform work for an agreed-upon price. But if there is no written contract or the written terms are ambiguous or conflicting, there is no agreed-upon scope of services for an agreed-upon price. Misunderstandings arise when the contractor's bid estimating does not include services that must be performed. The contractor's profit disappears as extras are rejected.

The second cause of poor contract drafting is the unwillingness of contractors to pay legal fees for something they believe they can do themselves. But even project managers who are aware of the importance of the need for proper contract documentation are not skilled in contract drafting and interpretation. Contractors may be familiar with standard contract forms that set forth their responsibilities and obligations. Standard forms are an excellent starting point, but they also have blank spaces for the parties to tailor the contract to the project. The devil is in the details. Standard forms sometimes lull contractors into thinking the details have already been taken care of and that contract formation is not important. Contractors typically do not include legal fees when estimating their bids. This is a mistake. Projects and local laws vary. An experienced construction law attorney can review the proposed contract and explain its language in lay terms so the contractor understands his or her legal responsibilities as the job progresses. The attorney can point out and clear up any ambiguities. The attorney can set up a separate contract file for the contractor at the inception of the project with subfolders and reminders. As the job progresses, the attorney can give general advice about how to word correspondence, how to document the job file, and when to submit notices of claims. The experienced construction attorney can provide these services on a prearranged retainer agreement. This cost can and should be built in as a cost of the project during the bidding stage.

The second most common mistake in construction litigation is making—and accepting—a bid that is far too low. Estimating is an art that should be based on reality and experience. When a contractor bids with little past experience in the type of project, using standard formulas and measurements, the tendency is to discount and hope for the best if the bid is accepted. Problems then develop when the contractor takes shortcuts to save costs because the project has been underbid. The contractor that is bidding in unfamiliar territory should hire an outside consultant with expertise in the type of project under consideration. The consultant can advise on areas where standard formulas may not be appropriate and can make suggestions for submitting a competitive bid that is based on experience.

The Importance of Listening

In order to be successful in this field, you must listen. Listening to the client is the first step toward understanding and successfully trying a case. Construction projects are often complex. The project manager or job site supervisor is usually a valuable source of the chronological sequence of events.

At the initial client meeting, sit down with the client and ask to be told about the job without reference to contracts or documentation of any kind. I strongly recommend against having a recording device present. Recording devices inhibit open communication because of the fear of confrontation by the contents of a recording in a later stage of the litigation. You want the client to open up to you, to pour out facts intermingled with feelings. You are not there to confront the client. You are there to learn why the client is embroiled in a predicament that requires your help. Put no time limit on the length of the discussion, and insist on no interruptions. Explain that the meeting is subject to attorney/client privilege. Take only notes that serve as reminders to ask appropriate follow-up questions. Let the person who put his or her body and soul into a project for several years pour out their feelings. Be a sounding board, but make no judgments. Gathering information in this manner allows you to develop a rapport based on trust.

Most people form conclusions about events without consciously understanding the facts in support of their conclusions. The "good facts" of the case will come first, and the "bad facts" will gradually work their way into the discussion. You may be able to explain that the bad facts are not as bad as they seem to your client or that additional facts are needed to evaluate the perceived bad facts. If a fact is perceived as good, probe its basis. For example, an architect may order that additional "minor" work be done by the contractor without additional payment to the contractor. The contractor may perceive that the work is not minor and that payment is owed for an "extra." It is necessary to understand the facts supporting the conclusion that the work is not minor. The contractor may believe as bad the fact that no protest was made when ordered to do the extra work. Probing the factual reasons behind the contractor's failure to protest may reveal that the scope of the work ordered to be done was insufficiently communicated to the contractor. Or the contractor may be pleased to learn the contractor has no legal or contractual obligation to protest when ordered to do additional minor work. Or a contractor may perceive as bad the failure to submit a claim for an extra within a contractual time period. Probing the facts may reveal that the contractor did not have sufficient information to recognize the need to submit the claim until a later time. The contractor can then be informed that this period of insufficient factual knowledge is a good fact because it extends the time during which the extra must be claimed.

During this process, you will begin to develop your theme. When the client understands you will fight for his or her cause to the utmost of your ability, your client will trust you, and your chances of achieving a successful resolution of your client's case will increase.

Every construction project is different, and each case is a learning experience. Therefore, it is important to understand the construction project from start to finish. If possible, visit a similar project with your client. As successful results are achieved, the hard work of combing through meeting minutes or change orders for facts in support of your client may become less tedious. Indeed, finding the missing note or drawing that supports the developed theme of the case is often exhilarating.

Preliminary Research

When first taking on a construction law case, it is important to obtain the arbitration claim documents or the complaint. These barebones allegations give notice of the contentions raised against your client. Next, obtain the complete contract documents in order to determine your client's role in the project. Whether you represent the owner, designer, general contractor, or one of the subcontractors, the nature of the project and the lines of authority are essential to understanding the causes of the disputes.

Obtaining this information enables you to begin the process of immersing yourself in the facts and law of the case. In this way, past cases or claims may come to mind that may be helpful in trying this new case.

As mentioned previously, my preliminary research always involves seeking a chronological review of the entire project from the client's first involvement. Obtaining this information is important, because although you constantly encourage and explore settlement, you must start with the assumption that the case will proceed to arbitration or trial. Depending on the forum, you may need to explain the project in very basic terms to a fact finder. Success depends upon the fact finder's understanding of your client's position. The better you understand what happened, the easier it will be to develop a theme and present that theme to an arbitration panel, judge, or jury.

At the same time, while it is essential to understand the facts from your client's perspective, it is equally important to understand them from the point of view of the other parties. The relevant legal principals must be analyzed. In theory, all parties are operating from the same understanding of the applicable law. An understanding of the entire case, from all viewpoints, will enable you to advise and consult with your client as the case progresses. In your initial client meeting, you start near the top of the organization's power structure. This may be a vice president or a high-level manager who has been assigned to manage the litigation. This individual has authority to make strategic decisions or has access to the decision-maker. He or she usually does not have first-hand knowledge of the facts and has reached conclusions based on information provided by others. The company's litigation manager should make it known to everyone in the organization that legal counsel is to receive their full cooperation. In subsequent meetings, you start toward the bottom of the company's power structure. These are individuals who provided the initial facts to the higher-ups. Often, these facts are filtered, so the boss has reached conclusions based on insufficient or incorrect data.

Understanding the Project

I have never had a client who was ambivalent about his or her case. Most often, the client starts with a desire for vindication. Occasionally, a client accepts blame and, too readily, fault.

The client's financial status has a strong impact on whether a case proceeds. Construction litigation is time-consuming and expensive. The amount of money at stake usually determines whether and how hard a case will be fought. Sometimes a company's very existence depends on a successful outcome of the litigation. Other times, whether a profit is made or a loss sustained on the project is determined by the resolution of the dispute. After the client understands the costs involved, I invest myself in giving my client the best representation possible. This requires total immersion in the history of the construction project. I take this approach because I believe every American trial lawyer owes his or her client their best—plus, I like to win.

I have found through experience that a thorough understanding of the project from a global viewpoint encourages creativity. Every construction project has multiple players with different expectations. Understanding what the owner thought the project was delivering for monies paid is essential. The architect's and/or engineer's design methods attempt to create on paper the owner's intent. The construction team has the daunting task of actually building what the designer thinks the owner thinks is being constructed.

When the client understands you are invested in his or her case, you gain trust. Initially, the client thinks you only took the case to earn a fee. When the client understands you really do want to win or obtain the best possible result under the circumstances, rapport develops. This approach is invariably successful, whoever the client is and whatever the type of construction dispute he or she is involved in. Nor does it matter what side of the dispute your client is on. Although they played different roles, all parties were involved in the same project. When you understand the entire project, your natural creativity awakens. Knowledge of other parties' roles enables you to understand their approach to the same set of facts.

Key Questions

In order to determine the best strategy in a construction matter, I must find the answers to three key questions:

- What is the client's response to the allegations made against them?
- What is the basis for the response?
- What evidence do we need to prove our case?

To that end, I generally ask each client involved in a construction dispute these questions:

- What is your response to each allegation?
- Who within your organization has the most knowledge about that issue?
- What documentation exists concerning that issue?
- Is there any publication I can consult to learn about that issue?
- Is there anyone outside your organization you can recommend who has sufficient knowledge to serve as a consultant to me on that issue?

These questions are asked for self-education. First, I have to understand the client's position and approach to the allegations made against them. Second, I need to sufficiently understand the issues in order to begin to develop a case theme. Third, I need to know how I am going to prove that theme.

The client's response to the allegations is the first step in analyzing the problem. Something happened during the course of the project to cause a dispute to ripen to the point where litigation counsel is necessary. Sometimes the client will have already invested substantial time and effort in resolving the dispute. Questions to consider include:

- How hardened is the client's position?
- Has the client made any harmful admissions?
- Does the client want to try to preserve this business relationship?
- How does the client appear to react emotionally to the allegations?
- Is an amicable resolution still possible?

Next, you need to make immediate contact with those individuals most knowledgeable about the issues. Has anyone left the company, and was it on good terms? Is someone who might be of assistance on a new assignment? Have key people been instructed to make time for conferences with legal counsel?

Based on your client's responses to your requests for information, documentation must be preserved and reviewed at the earliest opportunity. Outside sources, codes, standards, textbooks, and trade articles must be obtained and reviewed promptly. Outside consultants who are experts in the field must be retained at the earliest opportunity.

The Importance of Documentation

Construction projects are a multi-party undertaking. Understanding the roles of the participants and the lines of authority is essential. Documentation specific to the issues sheds light on how the problem developed and was approached by the parties during the course of the project.

All construction projects have a paper trail. Does one set of documentation lead to other documentation? Is documentation missing? What is the proper interpretation of the documentation that has been produced? Did the client make any admissions of fault? Did any other party make any admission of fault? Is anyone named in the documentation who may have key knowledge about the issues?

In order to answer these questions, it is essential that all project files be retained in their entirety and that all contracts, notes, correspondence, and writings of any kind dealing with the specific issue(s) of the case are sent to legal counsel immediately.

The Client/Attorney Relationship

Trust is a key ingredient of the successful attorney/client relationship. Trust is developed when the client understands you are working for his or her interests, even though the client may have made mistakes during the course of the construction project. The concept of attorney/client privilege helps enable the attorney to gain the trust of the client.

Often, an attorney must give "hard advice" to a client. Indeed, it is the attorney's duty to make every attempt to steer the client toward the best strategy even when the client does not want to accept that strategy. The leaders of many organizations are often surrounded by individuals who are adept at telling the boss what the boss wants to hear. Legal counsel, on the other hand, must tell the boss what the boss needs to hear. If the boss trusts their legal counsel, the boss will listen. This does not guarantee your advice will be followed, but developing a relationship based on trust will allow you to communicate your advice so it is considered.

The Client's Motivation

Understanding the client's motivation is essential in terms of developing a successful litigation strategy. Is the client so invested emotionally in the dispute that it overrides good business sense? Does the client want to vindicate himself or herself regardless of the cost?

Every case has settlement potential that should be explored frequently. If the client is so emotionally involved that settlement will not be considered, the costs of vindication might cause the client to reconsider. I once represented an architect in a case in which the building's exterior was failing. The architect blamed the manufacturer of the siding. The manufacturer blamed the contractor for improper installation. The contractor blamed the architect for faulty specifications. During mediation, the mediator suggested a site inspection. During the site inspection, the architect overheard one of the attorneys whisper to his client that the building's design was "ugly." The architect's hurt feelings nearly prevented a settlement until I explained what the cost of full litigation would entail.

Understanding a client's emotional reaction to a case is most useful in determining the timing of giving advice. You may find that a client who wants to be vindicated has to first realize on his or her own terms that vindication will be very expensive or will not even be accomplished. Is the client's bottom-line profit the most important factor, or is preserving a business relationship worth eating some crow?

The Client's Goals: Ethical Considerations

The best way to determine your client's overall goals in construction litigation is through frank discussion. In some cases, however, the client may express a goal that is not actually desired. For example, while an architect may want to prove his or her specifications were not faulty, he or she still does not want to lose the client for future projects. You must explore these competing goals with the client. Sometimes individuals within an organization have different goals. When this is the case, it is necessary to have the client decide its ultimate goal.

Whatever the client's goal, you must support it if at all possible. If you cannot, you must withdraw as counsel. In rare cases, your client may want to hide or even destroy evidence. This must never be condoned. I explain that anything unethical, illegal, or "against the rules" is out of the question. On those rare occasions when the client insists, you must withdraw as counsel. In other situations, I explain in detail why I disagree with proposed strategy. I tell the client that if he or she is not satisfied with my representation, the client has the right to discharge me. If that occurs, I explain that I will work with new counsel to the extent necessary to bring the new attorney up to speed. If I disagree with the client insisting on a course of action that is not dishonest or unethical, I confirm in writing my reasons for disagreement but confirm my commitment to the client's cause.

If you have properly invested yourself in a case, you are unhappy when you come to the conclusion that your client's goal cannot be obtained. When this happens, you must express to the client your dissatisfaction that his or her goal cannot, in your opinion, be obtained. You should review the facts of the case and the law in order to explain why you are disappointed, and then ask the client to offer suggestions on how next to proceed.

In helping the client formulate his or her goals, I always ask the client to try to view the case as though he or she were an independent arbitrator who was asked to decide the case. I then present the case to the client from all sides. Next, I ask the client how he or she feels about the case. This method usually results in the client reaching objective conclusions about the merits of his or her position.

Doing this type of analysis is important, because in many cases the parties to a construction dispute that has reached litigation do not fully appreciate the opposing party's positions. Each party, owner, design professional, or contractor may understand their own role, but they have not been trained to comprehend the role of the other parties.

Developing Case Theory

I consider case theory the theme of the case, and the simpler it is, the better. From a layperson's perspective, case theory boils down to just one question: What happened?

The non-lawyer client is usually helpful in developing a theme for the case because of his or her perspective. In describing his or her case, the owner may say, "I paid for a building that would function. I hired an architect and a contractor. Now they won't stand behind their work." The architect may say, "I specified using the best equipment, but the owner wanted to cut costs and didn't perform needed maintenance." The contractor may say, "I installed the equipment according to their plans and specifications, and now they won't pay me." These statements, which do not use legalese, are a good basis for developing a case theme.

The fact finder, be it an arbitrator or a jury, will be influenced by emotion. A statement such as, "This is a wrong that needs correction! We will prove that ABC Construction has been wrongfully deprived of its profit," makes a far more persuasive theme than, "The evidence will show that the defendant breached its contract with ABC Construction."

Establishing Boundaries

It is essential to establish certain boundaries with your client in a construction matter. At all times, you must be ethical. The client must know you will never hide or distort evidence or misrepresent the facts or the evidence to the opposing parties or the court. The same is true of the law. If the law is unfavorable, you must disclose it to the court or arbitrator at the appropriate time.

Establishing such boundaries is important, because it is the foundation on which our adversary system of legal representation is based, and it is morally appropriate. If the client does not respect these boundaries, you cannot represent the client. While you are willing to invest yourself in the representation of your client, if the client requests unethical representation, you will never be able to develop a true attorney/client relationship. You will not trust the client, and the client will not trust you.

Achieving Settlement

The client's wishes, within the bounds of ethical behavior, must be followed at all times. Once the client understands the costs and provides an adequate retainer, I proceed with litigation full speed ahead. On the other hand, if the client wants a quick settlement and gives authority to settle, I settle the case.

If the client wants to settle at all costs, the case is quickly settled. If the client wants to settle but only at a given figure, the circumstances are more difficult. As litigation progresses, costs increase. A realistic case budget must be submitted early in the litigation so the client can assess the costs of litigation as opposed to the costs of settlement. This process is ongoing as the case progresses.

Clients with no previous experience usually do not want to settle. On the other hand, clients with previous experience in litigation usually want to settle because they are more likely to understand that a "win" is almost never a certainty, that most cases ultimately settle anyway, that the only real winners in litigation are the lawyers who make a fee, and that preserving business relationships can be more valuable than winning a particular dispute.

Dealing with "Bad Facts"

"Bad facts" (i.e., negative facts) regarding the client's side of a construction dispute must be learned as soon as possible, and the impact on the ultimate outcome must be analyzed immediately. Legal research is often necessary to determine whether the facts are as bad as they seem or whether they can be deemed not relevant or negated by some other facts.

If the bad facts can be minimized or explained away, when presenting the case to a jury or an arbitrator I tell the fact finder in my opening statement what they are going to hear and then explain why the fact is not important, relevant, or decisive. For example, I once had a case in which a contractor had admitted fault to an owner. Usually, such an admission would be the end of the case. However, the early admission of fault was based on an inadequate factual investigation and an effort at an early compromise that would preserve the business relationship. Thus, I explained in my opening statement that the admission was based on a false factual assumption. I then explained what would be presented as the "true facts" and asked the jury not to make the same mistake as my client and reach conclusions on a false set of facts. Fortunately, it did not.

Ultimately, the client must always consider whether it wants to settle. If not, a strategy can always be developed to deal with any bad facts. The strategy may be as simple as "posturing." That is, being seemingly unconcerned at the mention of the bad fact. This does not mean the fact finder will also decide to be unconcerned. Some facts are, indeed, insurmountable. When this occurs, confirming your advice to settle in writing is essential. It may be that your client just wants his or her day in court and is willing to pay for that. As long as you are proceeding ethically and have fully informed your client of the risks, it is your ethical duty to follow your client's instructions.

The Importance of Alternatives and Non-Legal Ramifications

When developing a winning strategy for your client's construction dispute, you must always remain open to alternatives. Factual investigation or a change in the law can sometimes impact an entire strategy. Even when the overall strategy is to proceed to a final verdict or decision, settlement should be discussed. Newly learned facts or a change in the law can result in the development of a new strategy.

Non-legal ramifications must be considered and discussed with the client in the early stages of the litigation, because a "win" in construction litigation is not always a win. Can a client or business relationship be preserved? Will the litigation have an adverse impact on the client's reputation, win or lose? Is it better for the organization to put this project in memory and look forward? These considerations are part of the cost of litigation and must be considered and evaluated by the client at each stage of the proceedings.

Staying on Course

Ensuring that the course your client wants is actually followed is an ongoing process that requires evaluation from the initial meeting and during the case progression. In-person meetings or phone conversations, with follow-up confirming correspondence, is essential.

In all cases, counsel and client must have a true understanding of the facts. What happened, when, how, and why? Counsel must then advise the client on the applicable law. The other parties to the dispute have taken their positions because of their interpretation of the facts and the law. Who is right? Ultimately, the facts and the law control the outcome of the case.

When Strategies Go Wrong

Following the wrong strategy usually leads to losing the case for more dollars than what the case could have settled for at an earlier stage of the litigation. If you do not understand the client's motives and desires, you are probably working toward the wrong goal. This will impact the ultimate outcome of the case. The client will not be satisfied with the result and will likely blame counsel.

Fortunately, this situation can be avoided through continual attorney/client communication from case inception to closing. When problems with your strategy are encountered, backtrack. Try to determine at what point and under what assumed set of facts the strategy was set. Determine at what point you understood the client's goals to be different, and attempt to reach an agreement on what was understood between you and your client, and at what stage. Next, try to reach a new agreed-upon strategy or goal.

Outside Factors

Several outside factors can impact the strategy you develop for a client in a construction matter: opposing counsel, the forum, the relationship among the parties, and the availability of expert consultants to assist you and your client.

The quality of your opposing counsel—and your knowledge of them—can be very important. Sometimes you can reach agreement with counsel based on mutual respect. Other times, you can determine that opposing parties are taking untenable positions due to their counsel's advice. Competent opposing counsel should be respected for their competence. Questions to consider include:

- Will the opposing party's lawyer be more likely to persuade the fact finder of his or her client's position?
- Why has he or she taken the case?
- What does he or she know about the case that I am missing?

The forum your case is heard in also determines how you will approach strategy. Is this matter too complex for a jury, or will a jury be overly sympathetic? What about the judge? Does the judge have experience in construction litigation? Have you researched the judge's published opinions for other construction-related cases involving similar issues? Has your or your firm's experience with the judge been positive or negative? Do you want to be in arbitration with no right to appeal, and do you have a choice? Does the contract require arbitration? What do you know about the arbitrator? Is he or she an experienced construction litigator? Does he or she usually represent owners, design professionals, or contractors? Has the arbitrator been published? Are there any reported cases in which the arbitrator is listed as counsel?

Sometimes parties who are on speaking terms can resolve disputes without the assistance of counsel. Even after litigation has begun, disputing parties can get together and settle cases in the interest of maintaining business relationships and saving lawyers' fees.

When considering outside consultants, determine if one or more qualified expert witnesses will support your client's position. Does the expert have experience in similar or identical issues? Have you or your firm had previous experience with the expert? Is the expert published? Is the expert a "jack of all trades but master of none"? Can he or she assist in case investigation or preparation? What will it cost to hire an expert? Will the expert give an honest assessment? Will the expert be able to explain his or her opinions in a non-technical manner that is understandable to a layperson without experience in his or her field?

A Common Mistake

In my experience, a construction dispute proceeds to arbitration or trial only when I or the opposing counsel have improperly evaluated the case. Sometimes this occurs because of a mistaken interpretation of the law. More often, it occurs because the legal counsel does not sufficiently understand all sides of the dispute. Attorneys are often blinded by their client's position and fail to see that the opposing side's arguments have merit. The failure to recognize the merit of the opponent's position often results in giving the wrong advice to the client.

In order to avoid making this critical error, I always advise attorneys to keep an open mind. Constantly ask yourself and your client: What are we missing? Why are they (the other side) taking this position?

It is also essential to remind your client that settlement is almost always preferable to a verdict. The results of litigation are unpredictable, and even a win can be lost on appeal.

Jonathan D. Herbst was admitted to the Pennsylvania Bar in 1970. He received a B.A in 1967 from Pennsylvania State University and his J.D. from the Dickinson School of Law in 1970. He is a member of the Philadelphia (civil litigation section) and American (torts and insurance section on litigation) Bar Associations, the Defense Research Institute, and the Philadelphia Association of Defense Counsel. Mr. Herbst has represented design professionals and contractors in numerous jury trials and arbitration proceedings.

Dedication: *To my beloved wife, Sharon.*

Defense of Claims with Insurance Implications

Joseph E. Cavasinni, Esq.

Chair

Reminger & Reminger Co. LPA

Defining the Role of a Construction Lawyer

I view the task of a construction lawyer as being no different from that of any other lawyer. I am charged with the responsibility of solving my client's legal problems. In order to solve a client's problems, it is necessary to understand the cause and nature of those problems, understand the potential legal ramifications (which is often equated with liability and damage exposure), and develop a strategy to resolve those problems in a manner that is the most cost-effective for the client. Depending on the circumstances of the given situation, a client's legal problems can be resolved either pre-suit or after suit has been filed. However, even though the role of a construction lawyer is similar in a broad context to any other lawyer, a construction lawyer's role is different in that the claims in the construction context are not always just about money. Instead, the focus in many cases is on how to remedy or replace allegedly defective work.

My practice is exclusively devoted to litigation of construction claims, which includes lawsuits filed in state or federal courts as well as demands for arbitration filed with the American Arbitration Association or other alternative dispute resolution forums. I have represented owners, contractors, subcontractors, architects, engineers, and material suppliers in a variety of construction-related matters including tort claims (i.e., negligence) involving personal injury, death, or property damage; damage to the constructed product itself (i.e., construction defect claims); claims for delay and change order costs; mechanics' liens; bonds; and contract claims.

In many instances, I am retained by an insurance carrier to defend a claim against an owner, contractor, subcontractor, architect, or engineer, which puts me in the position of having two clients: the insured party and its insurance carrier. This "dual-client" relationship can be tricky, especially when there are adversarial issues between the insured party and its carrier such as whether the claim is covered and, if so, what happens strategically if the claim amount exceeds available insurance. Ultimately, since my principal obligation is to defend the underlying claim against the insured, my primary obligation as counsel is to defend the claim against the insured and remain from getting involved in any adversarial issues between the insured and its carrier.

Finally, litigation of these claims is not restricted solely to an actual trial or arbitration. It also includes all pre-trial or pre-arbitration activities including the written discovery process, retention of expert witnesses, conducting fact and expert depositions, and mediations. For purposes of this chapter, I will focus on my role as a construction lawyer in the context of being retained by an insurance carrier to defend a claim against an owner, architect, engineer, or contractor.

Adding the Most Value for a Client

From the big-picture perspective, there are two areas where I feel I add the most value for a client. First, one of my fundamental roles is to objectively assess and analyze the risk of liability and damage exposure to my client in a given case and gain an understanding as to whether there is the risk of personal exposure to the client. I focus on these issues at the outset of a matter in order to develop a strategy for defending a matter based on those perceived risks and damage exposure instead of taking a cookie-cutter approach to handling the file. A customized approach is far more valuable to my clients than a template strategy that does not take into account the uniqueness of their circumstances.

Another area where I add value for the client relates to my obligation to advocate zealously on my client's behalf, which entails communicating my client's position or defense effectively to the other parties and the judge, jury, or arbitrators. Since many of the issues involving construction-related disputes are technical in nature and require the testimony of expert witnesses such as architects, engineers, schedulers, metallurgists, or petrographers, my goal is to simplify or present these issues in an intelligible way to the judge or jury since they typically do not have engineering or construction background and experience. In short, unless the judge, jury, or arbitrator understands my client's position and the rationale behind it, it would be difficult to fulfill my legal and ethical obligation to serve as a zealous advocate on behalf of my client, which is why the ability to translate and present information in an understandable manner is so critical.

The Components of Construction Law

In a broad sense, with respect to my practice I would divide construction law matters into two areas: payment claims and tort claims. Payment claims are typically brought as actions for breach of contract (or unjust enrichment in situations where there is no written contract) and will often be accompanied with a mechanics' lien or bond claim, because the bond or real property subject to the mechanics' lien operates as collateral to secure the underlying debt. Typically, the payment claims focus on the scope of a party's work under the contract, the extent of the work completed, whether there are back charges or other set-offs, and whether there are outstanding amounts owed.

Tort claims are claims for personal injury, death, damage to other property, damage to the building or structure itself, or delay or acceleration costs arising from a party's tortious conduct, meaning negligence, recklessness, or other intentional acts. Tort claims can arise in a variety of circumstances including injury or death to contractor employees working on the site; injury or death to third parties either during the course of construction or following construction as a result of collapse or failure of a building or structure; damage to adjacent real property (i.e., storm water runoff from a construction site that allegedly impacts a downstream property owner) or other personal property such as a leaking roof or building facade that causes damage to furnishings, equipment, or computers; or claims for delay/acceleration or change order costs against design professionals for allegedly defective plans and specifications.

I assist my clients with issues related to each component of construction law. With respect to the payment claims, representing my client typically entails ascertaining the scope of work of the contractor, subcontractor, or materials supplier advancing the payment claim, which is rarely black and white. It also involves an accounting of payments for the work and/or materials provided to the project and addressing any claimed back charges, deducts, or set-offs for work and/or materials that allegedly do not conform to the requirements of the contract, which is often asserted as a defense to payment. This is essentially done via written discovery and deposition testimony.

Tort claims are in large part expert-driven. For instance, in cases involving settling foundations, which are quite common, geotechnical engineers, structural engineers, and contractor experts will be retained to address issues such as the cause of the settlement, methods by which the structure can be remediated or repaired, and the cost of those repairs. Similarly, in claims involving storm water runoff from construction sites (i.e., claims that silt and/or sediment is getting into a downstream property owner's lake), the parties are likely to retain experts in the field of hydrology, civil engineering, aquatic biology, and limnology to address the overall impact of development on the volume of storm water runoff, the efficiency and adequacy of sediment and erosion control measures implemented during construction, and the impact (or lack thereof) to the downstream property owner's lake in terms of plant and fish life. For these reasons, I try to retain my experts as soon as possible in a given case in order to gain a full appreciation of the liability risk and damage exposure early on. In this way, I can tailor my strategy and file handling through the discovery process in order to best minimize the liability risk and damage exposure. Moreover, retention of an expert early on also gives me the opportunity to make sure there are no legal or technical issues that have not been raised or anticipated.

Historically, the longer a file languishes, the greater the attorneys' fees and expenses, which can be as high as several hundred thousand dollars depending on the number of parties, documents, and witnesses in a construction matter. For these reasons, instead of merely churning hours on a file—which would be a disservice to the client—it is best to formulate a file-handling strategy with an eye toward performing those file activities that will posture the case for resolution as soon as possible, whether by way of settlement, arbitration, or trial. This is typically accomplished through the retention of experts, written discovery process, fact and expert depositions, and filing any necessary motions with the court that may dispose of the case on its merits.

Financial Implications of Construction Law

In light of the fact that the majority of my practice deals with defending architects, engineers, contractors, subcontractors, or material suppliers through their insurance carriers, the most common area in which I see financial implications relates to situations where the insured party has no insurance, does not have enough insurance, or where the insurance carrier is denying coverage for one reason or another. This can lead to situations where the insured party is adverse to its own insurance carrier. For instance, if the damage exposure is greater than the available insurance, the insured party may try to push the insurance carrier to settle the case within policy limits in order to avoid exposing the insured party to personal liability above and beyond the limits of insurance. Alternatively, situations will arise where an insured party believes it has coverage for potential damages, whereas the insurance carrier may take the position that the claim is not covered. This is most common with respect to construction defect claims where the claim is that the building or structure itself is improperly constructed. Indeed, many insurance policies have a "work product" exclusion that precludes coverage in these situations. Although my ethical obligation in this dual-client relationship is to defend the insured party against the underlying claim and without putting myself in the middle of insurance coverage disputes, resolution of the insurance coverage disputes and the dynamics between the insured party and its insurance carrier can be a driving force when it comes to settlement, especially since many claimants focus on available insurance proceeds and look to recovery against the insured personally as a last resort. For obvious reasons, if a party has no insurance, is underinsured, or its insurance carrier denies coverage, this can have dramatic financial implications to the insured party.

On a smaller level, I should also point out that many architects, engineers, or other design professionals will have deductible provisions in their insurance policies whereby the insured party pays all defense fees and expenses up to the time the deductible is exhausted. This can have dramatic financial implications on the insured design professional, especially since the applicable deductibles generally range from $25,000 to $100,000, which makes it all the more important to evaluate the risk and exposure in a given matter as soon as possible to establish an efficient plan to move the matter to resolution by way of settlement, arbitration, or trial in the most cost-effective manner. This concept is further reinforced by the fact that many errors and omissions policies issued to design professionals have "declining limits" provisions whereby the available limits to pay a claim or judgment are reduced by defense fees and expenses.

Common Client Mistakes with Respect to Construction Law

One of the more common mistakes I see clients make is in situations where the insured party executes a written contract for the project that may compromise the availability of insurance coverage for the claim or obligate the insured party to defend or indemnify another party even though the insured party's insurance carrier would not be obligated to defend or indemnify that other party. As an example, since many states require privity of contract for certain types of claims, an owner with a claim arising from the sub-consultant's work will only have to assert a claim against the architect with whom it enjoys privity of contract. This leaves it up to the architect to pay the claim and then recover from the sub-consultant. However, a common clause in contracts between an architect and its design sub-consultant limits the architect's damages in a claim against the sub-consultant to the amount of that sub-consultant's fees. If, however, the architect's insurance carrier was in the position to pay a claim and then recoup some or all of that amount in subrogation from the sub-consultant, the insurance carrier for the architect may be able to deny payment to the owner on behalf of the architect (even though the claim would have otherwise been covered) if the waiver or limitation of damages prejudices its subrogation rights.

Similarly, a clause in the contract between an architect and sub-consultant may require the sub-consultant to defend and indemnify the architect from and against any and all claims arising from the sub-consultant's work. However, the scenario can arise where a claim may be the result of both the architect's and the sub-consultant's work, which could potentially obligate the sub-consultant to defend and indemnify the architect under the contract, even though the insurance carrier for the sub-consultant would not be obligated to defend and indemnify the architect on behalf of the sub-consultant since the claim also arose from the architect's conduct.

Ultimately, the interplay between contractual language regarding defense and indemnity obligations and the obligations an insurance carrier may have with respect to defense and indemnity can have dramatic implications that can leave a party with tremendous personal exposure. Because each state will likely have a body of case law and statutory law dealing with the validity of defense and indemnity provisions in construction contracts, it is important to retain counsel with knowledge of the law as to these issues before entering into contracts that could potentially leave an insured with tremendous personal exposure.

Another mistake my insured clients sometimes make is that they erroneously think I am the "insurance company attorney" instead of their attorney. Nothing could be further from the truth. Although I am retained by the insurance carrier to defend a claim against a particular insured, which gives rise to the dual-client relationship, ethically my primary duty is to defend the claim against the insured party. Thus, if a situation arose where the best interest in defending the insured party was adverse to what was in the best interest of the insurance carrier, my obligation is to do what is in the best interest of the insured party. Ultimately, my insured party client needs to know and understand that I am its attorney, not the insurance company attorney, even though, at the outset of the claim, I may have an existing relationship with the insurance carrier and no relationship with the insured party. I feel that by thinking of me as the insurance company attorney the insured party is sometimes less than candid about its exposure and facts giving rise to the claim, which only detracts from my ability to defend the claim.

Strategies to Help Clients Deal with Construction-Related Issues

As indicated above, the areas where I feel I have the most value for a client are in evaluating the liability risk and damage exposure and effectively communicating my client's position to the opposing party, judge, jury, or arbitration panel in a way that makes sense given the technical nature of construction claims. In light of these primary roles, my initial strategy in defending a matter is to gather as much historical information as possible regarding the project in addition to any information or documents concerning the substance and basis for the claimant's liability theory and claimed damages. In addition to meeting with the client and reviewing its file materials, this is often accomplished by way of informal written requests and calls to opposing counsel or non-parties as opposed to otherwise gathering that information through the written discovery or deposition process, which takes longer. I have found this approach results in a more effective defense of the claim because an earlier understanding of the liability risks, damage exposure, and technical issues allows me to develop a case strategy that best minimizes the liability and damage risk.

Achieving Success in the Role of a Construction Lawyer

Communication is perhaps the most critical component to achieving success in the role of a construction lawyer. I often tell my clients that "the facts are the facts," meaning they cannot be changed. Rather, they can only be dealt with as effectively as possible. Too often, I have found that the inclination of the client is to disclose only facts that benefit the defense of the case, while omitting problematic facts. Then, at some point down the road, these adverse facts surface unexpectedly, which may cause me to change my analysis or file-handling strategy. The best approach is for the client to be as candid as possible up front and to disclose all facts, including those that are problematic to the defense of the claim. In this way, my analysis as to liability and damage exposure more accurately reflects the actual risk, and a strategy and game plan can be developed that take those problematic facts into consideration, including ways to potentially diminish their significance.

Information to Obtain Prior to the First Meeting with a New Client

One of the first things I try to do when I get a new file in is to meet with the client as soon as possible so we can become familiar with one another, and so I can obtain as much information as possible regarding the underlying facts giving rise to the claim, damage exposure, and potential defenses to the claim. Therefore, typically there is not a lot of file activity before this initial meeting with the client. However, I will at least review the complaint giving rise to the claim and contact the claimant's attorney in order to gain information informally as to the basis of the underlying claim, theory of liability, and claimed damages. This allows me to assess the theory of liability against my client and evaluate whether my client is a target or peripheral defendant. It also enables me to determine whether there are other potential parties that were not named in the complaint. Having this information allows me to share with my client the other party's position as to liability and damages. This then allows me to obtain information and documents from the client and develop an effective strategy for defense of the claim.

During the initial meeting with a new client, in addition to discussing the basis of the claim and theory of liability against him or her, I will try to obtain as much information as possible from the client regarding the underlying facts giving rise to the claim as well as relevant project documentation including plans, specifications, or contractor documents. It is important to have this information as early as possible so it can be shared with other witnesses, including expert witnesses expected to testify at trial or arbitration. Constructive feedback can then be obtained from these other witnesses, putting me in a position to formulate defense strategy.

A General Approach to Representing a Client

Generally, my goal is to try to obtain as much information and as many documents as possible—and as soon as possible—regarding the underlying facts giving rise to a claim, the theories of liability against the various defendants, and the potential damage exposure. This allows for a more meaningful evaluation of the liability risk and damage exposure earlier on, along with the development of a file-handling strategy from the onset of the case.

I have found this approach to be helpful in the sense that it allows me to be proactive with respect to written discovery, retention of experts, and development of deposition testimony. This has the effect of putting the opposing counsel on his or her heels, and it forces opposing counsel to react to what I am doing, ultimately providing me with some degree of control with respect to the course of discovery and the litigation in general. Ultimately, I have found this leads to more positive results.

Although I primarily do defense work, my approach in a construction dispute is the same in situations where I represent the plaintiff or claimant. Regardless of which side I am representing, I will always try to be proactive during discovery in hopes that I can control the course and path of litigation by putting opposing parties in the position of having to react to my discovery efforts.

Questions to Ask of a Client to Determine the Best Strategy

In general, the questions asked of a client in an initial meeting will focus on the underlying facts giving rise to the dispute, the role of my client on the project, the roles of other parties on the project, the identity of witnesses, and potential defenses to liability and damages.

I generally ask these questions to gain as much information and documentation as possible to allow me to appreciate whether there is risk to the client, what the risk is to my client as compared to the other defendants, and the viability of the plaintiff's damages claim. The answers to these questions also enable me to determine the categories of experts that may need to be retained to defend the claim, prepare a budget of defense fees and expenses for the insurance carrier, and begin to develop an initial strategy for defending a claim.

This information is also critical for ascertaining the potential for early resolution of a claim. For instance, if anticipated defense fees, expenses, and expert costs are substantially greater than the ultimate exposure on the claim if my client were found to be liable, I can put the insurance carrier in the position of deciding whether it wants to take these "economics of litigation" factors into consideration in deciding whether to resolve the claim.

Finally, this information also allows me to assess whether there is personal exposure to the insured party, whether because there is no insurance or because the exposure on the claim is greater than the available limits of insurance. That way, I can let the insured know there is the threat of personal exposure, in which case the insured party will often retain personal counsel to monitor the claim in order to make sure the insured party's personal interests are protected.

Critical Documentation to Receive from a Client

Generally, in my initial meeting with the client, I try to obtain all relevant documentation concerning the claim, which typically includes plans, specifications, contracts, and project documentation relevant to the claim, such as correspondence, photographs, or reports. The plans and specifications are typically highly significant, since often issues will arise as to whether the contractor, subcontractor, or material supplier performed the work or provided materials in accordance with the requirements of the contract documents. Similarly, as to errors and omissions claims against design professionals, the focus is often on whether those plans and specifications met the applicable standard of care. Moreover, obtaining contracts between the owner, contractor, subcontractor, architect, or engineer is also imperative, because many contracts will contain damage limitation clauses or provisions requiring one party to defend or indemnify another party from and against any and all claims arising from the first party's work on the project. As a result, one party may be required to pay the attorneys' fees, litigation expenses, and damages on behalf of another party. This documentation puts me in a position to fully appreciate the liability risk and exposure to my client. It also allows me to provide an anticipated budget of fees and expenses for the duration of the claim in order to allow the insurance carrier to set appropriate reserves.

Establishing a Positive Client Relationship in a Construction Matter

When it comes to establishing a positive client relationship, communication is a critical component. I am most effective in defending a claim against a client when that client is fully candid in that he or she has disclosed and provided all relevant information and documents, regardless of whether they positively or negatively impact the defense of the claim. Without that candor on behalf of the client, I cannot be effective. For obvious reasons, the attorney/client relationship impacts the situation in the sense that any discussions or information shared with the client, whether good or bad, are protected from disclosure to other parties. Moreover, this places me in a position of trust with the client and allows us to work together in defense of the claim and establish a strategy.

I always encourage a client to disclose all of the facts, good or bad. As mentioned, it is important for my client to understand that, while the facts cannot be changed, they can be dealt with, and by knowing about "bad facts" that are out there, I can anticipate how I will deal with those facts strategically in the event that they become public knowledge during the discovery process. Often, especially during trials or arbitrations, I will be the one to bring out the bad facts to a judge, jury, or arbitrator if I know they are going to come out anyway. That way, my client can generate credibility by acknowledging those facts that are not in its favor. Moreover, the impact of those facts can be minimized compared to the situation where they are brought out by opposing counsel.

Counseling Clients with Unrealistic Goals

Although I am ethically obligated to be a zealous advocate on my client's behalf, I would do a disservice to my client if I merely accepted my client's position on the matter without making an independent analysis as to the degree of risk and exposure to my client as to liability and damages. I cannot tell you how many times an insured client has indicated to me their belief that they did everything right and that the claim against them had little or no merit, only to conclude otherwise after independently evaluating all of the facts and learning the positions of the other parties. This is certainly not to suggest that I do not believe my client or am looking to find ways my client may be responsible. Instead, an independent assessment actually benefits the client since it puts me in position to fully appreciate the risk of exposure to my client and develop a strategy for minimizing or eliminating those risks.

As soon as practical in a case, I will draft a written status report to my client that sets forth my understanding of the underlying facts, the basis of the claim against my client, my assessment as to liability and damage exposure, and a discussion of issues that need to be resolved. I will also include recommended action items for further handling consistent with the defense strategy. This allows the client the opportunity to get an overview of the entire matter as I see it and provide input in case he or she sees the case differently or does not wish to proceed in the suggested manner. In these situations, if I believe the client's goals are unrealistic, I will simply tell the client, frankly and directly, so we can have an open discussion as to why we may see the case differently. This strategy is helpful since it allows any differences or misunderstandings to be resolved and puts my client and me in the position of working harmoniously to defend the matter.

In most situations where insurance applies, the decision of whether to settle the case falls with the insurance carrier instead of the insured. Thus, if the insurance carrier agrees with my analysis as to the risk of exposure to the insured party, the insurance carrier can still resolve the claim notwithstanding the insured party's objections. In situations where it is a "consent" policy, as is the case with many errors and omissions claims against architects and engineers, I will leave it up to the insurance carrier and the insured party to decide whether to settle the matter based on my recommendations. In situations where the insurance carrier may disagree with my analysis and recommendations, the ultimate decision rests with the insurance carrier as to whether to settle the claim or proceed in a particular fashion.

Developing the Theory of the Case in a Construction Dispute

The development of a theory of the case in a construction dispute is often a collaborative effort involving me, my client, and any experts based on an analysis of relevant documents and information concerning the basis of the plaintiff's claim, theory of liability, and claimed damages. I think it is extremely important to have the client, counsel, and experts all on the same page to promote consistency in the defense.

I generally try to not establish boundaries with a client when defending a construction matter. In short, I want to be freely accessible to my client and vice versa, and I want my client and I to be in a position where we can be candid with one another and say what is on our respective minds so any potential differences can be resolved. This enables us to work together and coordinate a defense of the claim. For these reasons, my clients know they can contact me at all hours of the day and on weekends, and that they are free to discuss with me anything that is on their mind.

My client's attitude in a construction dispute generally does not impact the voracity with which I proceed. Ultimately, my legal and ethical obligation is to do my best to defend a particular client, regardless of whether my client has a positive, negative, or indifferent attitude to the litigation.

How a Client's Willingness to Settle Impacts the Strategy

As part of my early evaluation of a claim, I make the determination as to whether a case can be settled at an early juncture for strategic reasons. If I make the determination that a claim is ripe for settlement, I will suggest that to the client, along with recommended settlement authority. If my client agrees with my recommendations, I will engage in settlement discussions with opposing counsel. However, in no circumstances do I want the plaintiff's counsel to think I am anxious to settle, which is why I will still undertake discovery efforts. Going through this process also prevents me from being caught off guard in case the matter does not settle.

I will say, however, that clients in construction matters generally do settle more often as compared to other litigation matters. One of the principal reasons for this is the fact that construction claims typically involve many parties, numerous witnesses, and voluminous documents generated contemporaneously during the project, which results in significant defense fees and expenses compared to other claims. For these reasons, the economics of litigation and cost of defense tend to be a greater factor in the client's decision as to whether to settle than in other types of legal situations. Moreover, many construction contracts require mediation of some sort before a claim can be litigated by way of a lawsuit or arbitration. This puts the parties in the position of having to come together in the spirit of compromise, which is obviously beneficial to resolution of the case.

The Impact of a Client's Financial Status on the Strategy

In construction matters with multiple parties, the fight is often not about who is at fault, but instead who is collectible. In other words, a plaintiff's attorney may change his or her strategy and may make a particular defendant more of a target because a party that is more at risk may be uncollectible. Although unfair, this happens routinely, which is why I will make the effort during discovery to understand what other parties to the dispute have insurance or, if not, whether they have the financial wherewithal to pay a judgment.

One concept that comes into play quite often is the status of a given state's law concerning joint and several liability. In simplest terms, in a tort (i.e., negligence) claim against several defendants, the jury will decide whether each defendant was negligent, and if so, the jury will assign percentages of negligence to each of the responsible parties. In a "several liability" jurisdiction, as a general rule, each defendant would only pay its proportionate share of the common liability. Thus, the plaintiff's counsel will often focus the claims and arguments on the parties that are collectible in order to have a greater percentage of the common liability apportioned to them. In a "joint and several liability" jurisdiction, however, each defendant is liable to the plaintiff for the entire award of damages instead of their proportionate share. Thus, a defendant that is only 1 percent responsible and has adequate insurance can get stuck paying the entire amount if the party that is 99 percent responsible is uncollectible. This would then leave the party paying the entire amount with a "contribution action" against the uncollectible party to recover the 99 percent the collectible party paid beyond its share of the common liability.

Clients Who Want to Follow Unadvisable or Unethical Strategies

If my client wants to follow a strategy that is not in his or her best interest, I will communicate that to the client but I will defer to him or her since the ultimate decision as to how to proceed ultimately rests with the client. However, if my client wishes to proceed in a manner that is unethical, I will advise my client as to why it is unethical and inform him or her that I am bound by the Code of Professional Responsibility to not proceed in that fashion. If my client insists we proceed in an unethical fashion to the point where I can no longer be effective in my representation of that client, I will file a motion to withdraw with the court, citing irreconcilable differences.

When presenting options for proceeding to a client, occasionally I find the client is ambivalent to his or her own construction dispute. Once again, my primary role is to evaluate the risk of liability and damage exposure to my client and make recommendations to proceed based upon that assessment. In this process, I will always present different scenarios to my client as to how we can proceed depending on the objectives of that particular client. In situations where there is some ambivalence on the part of my client, I will stress to the client that I can only evaluate the matter and make recommendations: the ultimate decision as to how to proceed rests with the client and needs to be made by that client. Thus, I use a frank and honest assessment of the situation, complete with recommendations, to combat ambivalence. If the client is non-responsive, there is little I can do, as the ultimate choice must rest with him or her.

The Most Important Aspect of Developing and Implementing a Proper Strategy in a Construction Matter

The most important aspect of developing and implementing a proper strategy is to make sure I have as much accurate information as possible as to the underlying facts giving rise to the claim, the role of my client in the dispute, the roles of the other defendants, any potential damages, and an appreciation of the technical issues that will require experts to be retained. This ultimately leads to the development of an effective defense strategy.

It is also essential to understand the client's true motives and desires in a construction dispute. Obviously, if there is no accurate understanding of the underlying facts, theories of liability, or the role of your client in the dispute, you run the risk of developing a defense strategy that is ineffective and a waste of time and effort.

Mistakes Made When Developing and Implementing a Strategy

One of the biggest mistakes I see attorneys make on a regular basis is allowing their assessment of the liability and damage exposure in a particular case to be skewed by information or documents obtained from their client, as opposed to evaluating information and documents from all sources, including retained experts. Sometimes a client will not disclose all facts to you, including those that are adverse to the defense. Moreover, a client may sometimes have an overly optimistic view of his or her position as to liability. This can result in an attorney forming an unrealistic assessment of the liability risk and damage exposure to his or her particular client. For these reasons, I believe it is imperative to obtain as much information and documentation as possible concerning your client's potential liability from other sources and address any inconsistencies with your client since, in the long run, your task is to identify objectively the liability risk and damage exposure rather than relying solely on the subjective analysis of your client.

Joseph E. Cavasinni is a partner with the law firm of Reminger & Reminger in Cleveland, Ohio, and head of the firm's construction department. He received a B.S. in civil engineering, with honors, from the University of Dayton in 1987. Thereafter, he worked in the road and bridge construction industry for eight years, including work as a project superintendent for a heavy highway general contractor and as owner of a consulting company specializing in polymer restoration materials. In 1995, he obtained his J.D. degree, cum laude, from Capital University Law School. Since he began practicing as an attorney in 1995, Mr. Cavasinni's practice has been devoted nearly exclusively to construction law, with a particular emphasis in construction litigation. He has represented owners, contractors, subcontractors, architects, engineers, and material suppliers in a variety of construction-related matters including tort claims involving personal injury and property damages, mechanics' liens and bond claims, and other contract claims. He is a member of the Ohio State Bar, the Ohio State Bar Association, the Cleveland Bar Association, and is licensed to practice in the U.S. District Court for the Northern District of Ohio.

Keys to Effectively Serving the Construction Industry

Calvin R. Stead
Partner
Borton Petrini LLP

The Duties of a Construction Lawyer

A construction lawyer handles virtually all phases of the needs of the construction industry. The role involves representing builders, developers, and tradesmen in every aspect of the construction process, beginning with the initial purchase of land and including the acquisition of entitlements and the development of a property. The duties of a construction lawyer encompass representation of the client in subsequent litigation and contract disputes at any point during the construction process. As such, construction lawyers need to be familiar with much more than contracts and the purchase and sale of property. Adequate representation includes the environmental document process and negotiating with public agencies as well as the transactional elements of any given development. Another essential aspect to the construction lawyer's skills is effective litigation involving each step of the development process.

As construction lawyers, our primary focus is on understanding the nature of the facts in any representation. Without a detailed understanding of the manner of production of the particular product at issue, it is virtually impossible to develop a strategy that optimizes the eventual outcome in favor of the client. By approaching each construction problem from this perspective, regardless of the nature of the trade or the specific client being represented, the representation proceeds naturally from the initial mindset of the handling counsel.

The Major Components of Construction-Related Law

Our primary areas of representation are developing and defending environmental documents, defending construction defect litigation, and providing counsel for individual builders and the industry as a whole on matters as diverse as contractor's liens, negotiating with city and county planning departments, drafting environmental impact reports, drafting contracts, advising on insurance matters, and handling litigation related to all these areas.

Construction law truly consists of "cradle-to-grave" representation of developers, builders, tradesmen, and associated licensed professionals such as architects and engineers with respect to identifying and purchasing land, developing the property in compliance with the Subdivision Map Act and environmental requirements, building to established construction standards, drafting related contracts, and defending the client or prosecuting cases on the client's behalf when litigation arises.

There are many value-added services we provide clients across the entire spectrum of construction law. We work closely with clients with respect to property acquisition. In addition to drafting and reviewing purchase and sale agreements for clients, we find they often need advice regarding formation on community service districts, community facility districts, and community service associations. Clients also require assistance with specialized title matters such as negotiating mineral rights waivers and securing water rights. Easements, licenses, and rights of way are often required as well.

With respect to environmental requirements, as lead counsel we work in conjunction with the development team to assist in drafting environmental documents and review the work of other team members, ensuring compliance with environmental laws. Often, specific questions come up with regard to the Subdivision Map Act, water supplies that comply with state requirements, and mitigation measures public agencies or environmental groups want incorporated into a project. Other components of our role in this arena are to help address and resolve these questions and represent the client's project before planning commissions, planning departments, and elected officials. A balanced sensitivity to political necessity is also required.

While actual construction of the project is indeed a component of the process, it is not typically a phase in which the lawyer is intimately involved. However, issues continually arise in the form of disputes between contractors and subcontractors, homeowner disputes, and compliance with statutes, codes, ordinances, regulations, and generally accepted construction standards. One example of this is when excess release of fugitive dust in the preparation of roads and building pads shut down a project, which will require immediate legal intervention. Another common problem is handling employee disputes for the client.

Finally, litigation can arise at virtually each phase of the construction process. The most common sources of litigation are breaches of contract, environmental challenges to the project, and construction defect litigation filed by disgruntled homeowners and owners of commercial buildings. Given the likelihood of such occurrences, clients are wise to retain an attorney with deep expertise in all facets of construction law to work on their behalf.

The Financial Implications of Construction Law

The construction lawyer does more than just prosecute and defend lawsuits or negotiate deals, although those are essential elements of the construction lawyer's arsenal. The economic value of advice to clients from a construction lawyer can be invaluable in avoiding problems or protecting against future lawsuits. For example, the transactional lawyer who drafts contracts protecting the client from unforeseen events can be of great benefit to any construction deal, providing value that is far greater than the cost of the lawyer's time. Often, simply reviewing a contract before it is executed will result in subtle changes in contract language, tailoring the document to the specific deal in such a way that the client will get substantially more out of the deal than otherwise would have been the case.

Similarly, the value of using a lawyer to shepherd environmental documents through the approval process assures the client of a more airtight document than otherwise would have been achieved. A properly prepared and processed document is capable of withstanding the onslaught of environmentalists bent on blocking a project. Many projects that otherwise might never have been built have been saved by creative lawyering and careful attention to details in the development process, ensuring that the legal details are addressed in such a way that the document is virtually impervious to onslaught from a legal perspective. Furthermore, city and county planners and elected officials are much more likely to approve a project that is legally sound than one that has even the smallest of legal deficiencies.

Providing Value for Clients

Providing value for clients comes from multiple sources. One of the most important ways to provide value for clients is by providing contract drafting and review services. In many ways, construction contracts are form documents. However, each deal is different. Therefore, each contract needs to be reviewed with clients once the lawyer knows what their client wants. By meeting with the client and coming to a detailed understanding of the deal, the contract documents are tailored to fit the deal so the involved parties get what they want with a minimum of risk.

In addition to contract documents, a good construction lawyer also provides drafting and reviews of homeowner manuals and regular updates based on changes in the law and requirements of insurers. These services are not particularly expensive. Clients who regularly update their construction documents are well equipped to maximize profits and minimize legal issues and challenges.

The input on environmental documents from experienced construction lawyers provides a similar benefit to clients. The careful review of each element of the environmental impact report provides significant value for the client by minimizing the risk of legal challenges and helping to ensure that the public agency gets what it wants in terms of drafting measures to protect the immense amount of effort that goes into each project.

Common Client Mistakes with Respect to Construction Law

One of the common mistakes builders make is trying to handle their contracts themselves. Even when the builder pays a lawyer to draft a contract, the builder will often negotiate deals without consulting a lawyer. This frequently results in a contract that does not fit the project, which in turn increases the likelihood of problems if the deal goes sour.

Another problem that frequently surfaces is a builder who has a good contract document but fails to follow his or her own contract requirements, thus negating the value of the document substantially. For example, where the contract prohibits deviations from the specifications without written modifications (i.e., change orders), we often see builders fail to execute some or all of the change orders requested by the client. Then when it comes time to pay the bills, the client refuses to pay for work done without a change order. By failing to follow his or her own contract documents, the builder has virtually guaranteed trouble getting paid for some of the modifications, which often makes the difference between breaking even on a project and making the profit for which the builder originally bargained.

Similarly, developers often try to have their engineers handle environmental documentation without involving counsel. In states like California, where environmental litigation has become almost a matter of course, this is a virtual recipe for disaster. The small amount of money saved by using engineers rather than legal counsel is quickly offset by legal fees incurred later to defend a defective document. It is also vastly offset by the delays incurred in having to litigate a project or recirculate an environmental impact report and miss marketing opportunities such as favorable interest rates or a booming seller's market.

Common Lawyer Mistakes

The biggest mistake I see other attorneys make with regard to developing and implementing a client strategy is the lack of a clear understanding of the operable facts. This usually results from a lack of adequate communication between the client and lawyer.

Put the wishes of the client first, and treat them as you would want to be treated if you were the client.

Strategies to Help Clients with Construction-Related Issues

Throughout our years of experience in this industry, we have found that crafting development teams is key to successful environmental planning. We help the client in selecting appropriate team members, and we oversee each team member to keep his or her work product consistent with the applicable general and specific plan elements. In fact, keeping all the team members on schedule is probably one of the most important functions the lead counsel provides. Since every delay in the process represents additional expenses to the client, setting deadlines and sticking to them is invaluable. This is the only way to keep these massive projects on schedule.

Another essential part of helping the client handle construction-related issues is to communicate often and in detail with the client regarding the project. For us, this involves weekly conferences with team members and the client and making assignments to each team member to keep his or her part of the project on time. Part of this process may involve putting pressure on team members who are not keeping to their schedules. Once a deadline is set, it cannot be missed. The lawyer has to take the responsibility to make sure delays do not occur and to facilitate the process.

The lawyer's success in this role with respect to working with clients is dependent on the willingness of the client to communicate with the lawyer. The successful construction lawyer must keep lines of communication open and must communicate consistently with each client by phone, e-mail, correspondence, and in person. Contractors and developers are notoriously independent and overworked. It is hard for them to take time out of their days to talk to lawyers. In fact, it is counterintuitive for them to seek out a conversation for which they will later get billed. Lawyers who are sensitive to this are much more likely to get repeat business and achieve success in this field.

The other essential quality of a successful construction law practice is responsiveness. In this field, it is critical to respond immediately and decisively to client inquiries and to give their problems instant attention. In many ways, construction clients are like military generals. They expect to be listened to, and they expect their orders to be followed and respected. The lawyer who can ask, "How high?" when the client says, "Jump!" and yet still have the confidence and experience to tell builders when they are wrong will be the lawyer who earns clients' respect and repeat business.

Yet another component of remaining successful in this field is to constantly sharpen one's knowledge and keep on top of current and emerging cases and trends. We ask our construction lawyers to personally review advance sheets daily for new construction cases. Those that directly affect the industry as a whole or clients specifically are brought to the attention of the clients, and strategies are implemented to protect the clients from such changes in the law. We also keep up with new laws being pursued by the legislature that affect the construction industry. These too are brought to the attention of the clients and incorporated in our activities on behalf of clients to keep our representation current and relevant. We attend continuing legal education seminars on construction-related issues, and we try to stay active in construction industry organizations such as the National Association of Homebuilders, the state Building Industry Association, and the local chapter of the Builder's Exchange. We regularly attend annual functions of these groups. We produce a quarterly newsletter on construction law and write a monthly column in the local Building Industry Association newsletter. Making this commitment helps keep a lawyer current on aspects of the industry that are detail-oriented. In addition to scholarly research and involvement, we recommend maintaining a hands-on approach to your work. Try to do legal research for clients yourself instead of handing it off to associates. We also strive to have the partner in charge handle as much of a construction matter as possible. Although these activities may not be billable hours, they are important in that they help the responsible lawyer keep on top of the issues that are critical to our clients, allowing us to be proactive instead of reactive in our practice.

Preparation before the First Client Meeting

The first thing we like to do when we get a new client call is to spend some time on the phone with the prospective client discussing the specific reason he or she has called. Once we have a clear understanding of the problem, we review various legal treatises to familiarize ourselves with any particularities of the specific issues and attempt to formulate a handling strategy before we first meet with the client.

This information is important, because many of the legal questions that surface with new clients can be both complex and unusual. By familiarizing ourselves with the legal issues before meeting with the client, we are more prepared to evaluate the case and develop strategic options we can explore with the client at the first meeting.

By preparing in this way, we are able to anticipate most of the questions we need to have answered by the client in our first meeting. These questions include:

- What are the contract terms? Do you have copies of all contracts?
- When was the work completed? Was a notice of completion filed, and if so, when?
- Did anyone provide you with an additional insured endorsement, or did you provide anyone with one?
- Did anyone else work on the project other than you?
- Who were your insurers and for what years?
- What are you looking to accomplish in retaining us?

The contract terms dictate the issues that will be litigated. The dates of completion will indicate when the statute of limitations began to run. The identification of insurers will enable us to tender the defense of the case. The availability of additional insured endorsements indicates other parties to whom the defense can be tendered or to whom the client owes a duty to defend or indemnify. Others who worked on the project will indicate those parties also responsible for the damages alleged or who may have been responsible. Knowing what the client is looking for gives clues to motivation and expectations and possible client control problems we may need to address.

Additionally, if the problem is a complex legal one, we can avoid wasting the client's time at the first meeting and we are able to get more information that is relevant to the problem in less time. This is always important to clients in this industry, who have little time to spare.

The Initial Client Meeting

In the first client meeting, it is essential that the lawyer understand the facts of the case in detail. This means asking the client to bring in the entire document file on the particular case. We need contracts, insurance declaration pages, notices of completion, the entire job file for the project, any complaints about the client's work, evidence of any warranty work, and before and after photos of the work, if available. We also like to have the daily logs for the job and the foreman's journal, if available.

All of this material will be used to support the claim that the work was done properly and any defenses we may want to assert to the claims. Similarly, if the buyer is disputing all or a portion of the work, this information will help prove the case against the buyer.

By obtaining this information from the client at the beginning of the dispute, we can assert appropriate claims and defenses in the initial pleadings. We will also get an idea of where the case is going to go and whether any defenses such as the statute of limitations will get rid of the lawsuit.

The more detail we can sort out in the first meeting with the client, the more likely it is that we get started on the right foot with respect to the representation. This also provides us with a chance to ask the client detailed questions about the facts based on our initial research so we can obtain information the client might not otherwise tell us about this early in the case. Once we get the information we need, it is also helpful to ask the client in a very direct fashion what he or she is looking for in the representation. Finally, we try to call opposing counsel or the other party with the client in the room, giving the client the opportunity to listen to the conversation on the speaker. This tactic helps the client understand in a personal way the obstacles to resolution we have just discussed. It also gives the clients a feeling of security that we will be communicating important matters as they arise and involving them in the process as it progresses. We also like to think the clients want to see exactly what we are doing for them and thus have a right to personally observe our efforts.

Clients are usually fixated on certain facts when they first come in to see us. These are the facts that have caused them to come to a lawyer in the first place. However, in the overall scheme of things, the client's perceptions, although important, are often not pivotal as to which facts will control the probable outcomes in the case. By doing preliminary research before the first client meeting and by getting the client refocused on the relevant facts, we not only obtain early focus in the case, but we tend to get better client control, which positions us well to develop an effective client strategy.

Formulating a Client Strategy in a Construction Matter

Usually in construction cases, the legal issues are relatively self-evident. Helping the clients understand their options is key to developing a resolution strategy. This is because construction clients are always interested in the bottom line and are moving on to the next project before the issues surrounding the one at hand are even resolved. Most clients seek our help while upset and emotional, and this drives them to want to go to court to get what they think they deserve. There are many other ways to obtain favorable results, and helping the client understand these options early in the case reduces the likelihood of an upset client later.

The relationship between the client's expectations and the reasonableness of the damages claims is one of the best clues to client motivation. If the client is being reasonable, both as to outcomes and costs to get there, we feel like we can proceed in a systematic way without having to educate the client on the legal issues.

In representing contractors, we usually don't have client control problems because of our approach to setting up the representation, as discussed above. This question is more applicable to a buyer who is suing a builder. Often, homeowners are not sophisticated in legal matters and have very unrealistic expectations about values of their lawsuit and probable outcomes. However, the key to successful representation in either case is clear and consistent communication and competent handling of the file.

Presenting Options and Establishing Goals

The client will tell you what they want once they understand their options. Our job is to make sure we get the client a clear message as to what their options are and what the likely results will be.

We are usually very firm in our presentation of options, costs, and outcome discussions when we suspect a client is being unrealistic. Once we have presented in a clear and concise way the realities of the litigation process generally and the probabilities of the particular case, if the client still doesn't get it, we repeat the information until they understand. Sometimes we provide an opinion letter to the client if we suspect client control problems. If the client is simply not listening or trying to get us to say something to support their unrealistic expectations, we suggest they may want to seek a second opinion. The lawyer should be brutally honest with an unrealistic client. Some of them listen, and others go looking for another lawyer. The tail can't wag the dog in litigation. If the client isn't listening, we can't help them, and they are better off going elsewhere.

The first decision that needs to be made is whether the case is one that should be tried in court or settled out of court. This decision pivots on at least three elements:

1. The relative liabilities between the parties (i.e., the merits of each side's case)
2. The relative costs to go to court or settle
3. The client's risk tolerances

For example, in a case where the client has a high probability of losing, we only want to have a trial if the other side refuses to be reasonable in settlement negotiations. Factors that impact our strategy include costs, likelihood of success or failure, damage assessment or the amount the client stands to gain or lose, and the time we anticipate it will take to get a favorable result.

The Impact of the Client's Financial Status

Defending a lawsuit does not normally depend on a client's financial status. From a plaintiff's standpoint, construction defect litigation is contingency litigation and is not impacted by financial status. In those few matters where financial status is relevant, we do not take cases for clients that cannot afford to litigate, and we advise them to settle, forget litigation, or consider other options.

Remaining Open to Alternatives

A lawyer cannot get a good result for any client without remaining open to alternatives.

Settlement is always an option. However, as a case evolves, alternative strategies become relevant as more facts become known. Another factor we consider is any potential delay that might impact this predicted timeframe.

The client also needs to understand the cost of various options and the likelihood of a successful outcome.

Building a Strong Client Relationship

Construction clients need to trust their lawyer. By knowing what the important issues are to keep down expenses, defend or prosecute claims, and hasten resolution, this trust can be developed. Experience and expertise are very important to contractors since they too are in a service-oriented business. Gaining the trust of the client and giving the client the sense that the matter is in good hands is probably more important than trial expertise to our clients, since most cases settle.

Cooperation from our busy clients is very important. They are much more likely to assist in the defense if they trust we are doing everything possible to get a favorable outcome as quickly as we can and in a cost-effective way.

Good communication with the client and mutual trust and respect help get the case resolved as quickly and effectively as possible. If the law is against the client, he or she needs to know it early on in the process. If it looks likely that the client will probably win in litigation but that the process will cost more in fees and expenses than what the client will receive, this information needs to be communicated early in the representation. In addition to analyzing the legal positions of the parties, a damages assessment is essential to this analysis. If the client has brought in everything we need, we will develop a damages assessment in the initial meeting so it is clear what the client stands to receive or pay out if the matter involves litigation.

If the matter does not involve litigation, we need to try to give the client a clear idea of the time, expense, and other resources necessary to achieve his or her goals, as well as the likelihood of success. Clients who are experienced in development probably know approximately what it will cost and how long it will take. This type of client is usually looking for a political analysis of obstacles among elected officials and city or county staff that could increase the expense of the project or cause delays. An inexperienced developer, usually a farmer or large landowner, needs to be walked through the entire process, as they often don't understand what they are getting into.

A General Approach to Representing a Construction Client

Our general approach to all litigated construction matters is to try to settle the case as soon as possible. In development matters, settlement is not an issue. Our primary concern is keeping costs down, keeping the project on track, and avoiding legal challenges by being proactive. If we find out we are going in the wrong direction, we reverse course, explain to the client why we are changing strategy, and execute.

We adopt this general approach because settlement is always preferable to litigation. It costs the client less in attorneys' fees and allows the busy builder to get on with the next project without legal distractions. If the client is at fault, spending money on attorneys' fees is throwing good money after bad. If the client is not at fault, it will often cost much more in attorneys' fees than it will to settle a no liability case.

This strategy is effective because clients understand that when we are trying to settle a case, we are trying to save them time and money, and they appreciate this effort. They know attorneys' fees are hard to recover in litigation, and they understand that success in court is never guaranteed and costs a lot of money and time. By focusing the case on resolution early and often, we believe we are best serving our clients and maintaining client loyalty by trying to optimize the client's economic interests.

The approach will vary depending on the client and the type of construction dispute in which he or she is involved. Some construction clients want to project a hardball image that is not consistent with settlement of a high percentage of cases. Where the liabilities are favorable, we place those matters on a trial track. Of course, we make sure the client knows the relative costs of this strategy.

If a client wants us to do something unethical, we simply tell them we can't do what he or she wants and why, and we suggest he or she locate other counsel. If the client wants to proceed with unreasonable strategies or in a manner that is contrary to his or her best interest, we make sure they are clearly informed of our concerns in writing, and then we do what the client wants. Even if we don't agree with the client, he or she is still the boss.

In construction defect litigation, early settlement is often not possible. This is especially so when insurance companies are involved in the case, since the insurance companies will dictate the course of litigation. In disputes between builders and buyers or builders and subcontractors, we usually try to play hardball with the subcontractors and softball with the buyers.

If the client is in a dispute with a buyer, it is usually because the buyer has withheld payment. In this circumstance, we are looking to get the client paid as soon as possible without incurring substantial fees. We are not looking to sue the buyer unless the amounts owed to the client are at least ten times the anticipated attorneys' fees. In these cases, we do everything feasible to facilitate settlement.

If the client is being accused of construction defects by an individual buyer, we try to fix whatever the complaint is and settle any payment dispute. In this situation, we play hardball with the subcontractor who did the defective work and may have to sue the subcontractor for repair costs. If the dispute is with another contractor, we try to settle matters that are adverse and place matters with favorable liability on a trial track. Finally, if the client is being sued by multiple homeowners or a homeowner's association, we need to tender the matter to the carriers and cross-complain against the subcontractors for all fees, costs, and indemnity for damages incurred by the plaintiffs and the insurer.

For a risk-averse client, we always try to minimize the confrontations and promote early settlements. Confrontational clients are more likely to expend more on fees and more likely to get into a trial.

Ambivalence of the client usually indicates a lack of will to proceed in litigation. We should be exploring settlement or alternative dispute resolution options with these clients. Detailed and open discussions with the clients usually resolve ambivalence. Often, ambivalence is caused by lack of clarity of the legal options. Once the client understands what their options are, they will be able to make a reasoned decision.

Overall, I don't think construction cases settle at a significantly different rate from most other types of civil litigation. Contractors tend to be more sophisticated than casualty claimants and therefore are often able to avoid litigation, so many of their cases settle before we ever see them. Time is money to contractors.

Dealing with "Bad Facts"

When it comes to "bad facts," we are very up front with the client. We tell it like it is and spell out in writing what the problems with the case are so there is no question later when things don't turn out the way they may have originally hoped.

When faced with lots of bad facts, early settlement strategies are preferable. We currently feel mediation is very helpful in resolving unfavorable liability. We would rather see a client pay up front to settle a case than incur a lot of attorneys' fees trying to defend bad facts and then still have to pay the fair value of a claim. On the plaintiff side, we simply counsel against lawsuits. With these strategies, the client saves money on fees, reduced stress, quicker resolution, and less wasted time on a losing matter.

Dealing with Non-Legal Concerns

The primary non-legal concerns involve warranties, builder reputation, and litigation economics. Sometimes delays or impacts on other business activities of the client are also important.

In our experience, non-legal issues are not especially important to litigation strategy. Where the representation is one of a developer seeking entitlements, they are extremely important.

Where political or mitigation of impacts are concerned, the creativity of the development team is essential to success. Weekly strategy sessions are just about the only way to keep up with the evolution of issues.

Believe it or not, the client's family often becomes a factor in litigation. We need to be sensitive to such factors when they come to light. The impact of litigation on the client's other business is also important to consider in some cases. Litigation is a money drain and is time-consuming.

Careful handling of extraneous factors should not impact litigation negatively. However, they can lead to significant changes in direction.

Final Advice

Remember that part of the job is to make the client look good in the eyes of their superiors, family, and peers. Lawyers can get arrogant and need to remember their job is to provide a service to their employers. As fiduciaries, we need to take care of clients as if they are family.

Calvin R. Stead is a partner in the Bakersfield office of Borton Petrini LLP. His urban planning background includes an undergraduate degree in zoology with a minor in organic chemistry and a master's of science degree from the University of Texas. He has more than twelve years of experience in air pollution chemistry, nuclear power plant chemistry, and chemical toxicology. In 1986, he received his J.D. from the University of San Diego School of Law.

Mr. Stead's areas of legal expertise include realtor errors and omissions, commercial and environmental litigation, land use planning, water law, and environmental and health care issues. Within the area of toxic tort, he has handled a vast array of cases including vaccine reactions, asbestosis and cancer phobia, chemically induced asthma, pesticide and herbicide contamination, EMF and AIDS contamination, and phobia claims.

He has consulted on and litigated complex Comprehensive Environmental Response, Compensation, and Liability Act and Resource Conservation and Recovery Act cases. Major clients in the environmental area include ITT, Maxxam, Lederle Labs, and American Cyanamid. He has also counseled and advised health care providers regarding Stark and anti-kickback and fraud claims as well as malpractice and medical staff issues.

Mr. Stead belongs to many professional organizations representative of both his scientific and legal expertise. Among these are the American Society of Clinical Pathologists, the American Nuclear Society, the Defense Research Institute, the Defense Counsel Association, the International Air Waste Management Association, the State Bar of California, and the American Institute of Planning. In addition, he has written many articles for California Defense *magazine dealing with topics as diverse as pollution insurance and zoning and land use issues.*

Mr. Stead's areas of legal expertise include toxic tort litigation, realtor errors and omissions, commercial and environmental litigation, land use planning, construction defects and oil field litigation. He has represented builders, developers, and subcontractors on a wide variety of construction issues, including grading, soils, foundation, asphalt, concrete, flat work, tile, framing, floor coverings, roofing, masonry and stucco. Within the area of toxic torts, he has handled a vast array of cases, including toxic molds, vaccine reactions, asbestosis, cancer phobia, chemically induced asthma, pesticide and herbicide contamination, EMF, and AIDS contamination and phobia claims.

Mr. Stead belongs to many professional organizations representative of both his scientific and legal expertise. Among these are the American Society of Clinical Pathologists, American Nuclear Society, Defense Research Institute, Defense Counsel Association, International Air Waste Management Association, the State Bar of California, and the American Institute of Planning. In addition, he has written many articles dealing with topics as diverse as pollution insurance and zoning and land use issues.

Mr. Stead stays as involved in his community as he is with his profession. Some of his community service activities include membership in the Bakersfield Rotary Club, membership on the Greater Bakersfield Government Review Council, and serving on the board of directors of Project Clean Air. He was the 1996 president of the Sonoma County Taxpayers Association and is now on the board of directors of Kern Tax. He is also a member of the board of directors of the Building Industry Association in Kern County.

Mr. Stead is married to Elizabeth and has two children, Christina Nichole and John Frederick.

A Focus on Representing Owners

Donald B. Brenner

Shareholder and Chairman, Construction Litigation Group

Stark & Stark

In my role as a construction litigator, I typically represent community groups such as condominium or homeowner associations as well as owners of commercial or professional buildings who have discovered serious deficiencies in the design and/or construction of their buildings and other property. Our cases involve all kinds of construction deficiencies and design defects including those relating to many different kinds of exterior cladding, roofs, sea walls, filigree concrete slabs, fire suppression systems, deck and balcony design and construction, windows, sanitary sewer systems, pump stations, septic systems, drainage issues, mold, electrical, HVAC, and mechanical. We also deal with claims relating to defective products used in construction. The most common defective product claims we make relate to exterior insulation and finish systems (EIFS) and thin brick systems.

EIFS is a synthetic stucco product that was originally designed to act as a face-sealed barrier to water penetration. It is therefore known as "barrier EIFS." The defect in barrier EIFS is that the system depends upon caulk joints at all windows, doors, and other penetrations through the system and where there is an interface between the EIFS and a non-EIFS surface such as a deck or roof. There is no secondary drainage system to allow any incidental moisture that gets behind the EIFS to escape. This frequently causes significant and sometimes catastrophic damage to structural framing and sheathing and facilitates the growth of mold. There have been hundreds of lawsuits and class action cases involving EIFS across the United States. By 1997, EIFS manufacturers began coming out with drainable EIFS products to replace their barrier EIFS. Eventually, barrier EIFS was outlawed in several states for use in one- and two-family residential construction.

Thin brick is a half- to 1-inch-thick brick veneer that is glued to an extruded polystyrene board that is adhered to a plastic track that is attached to the metal framing of the building. The thin brick systems we have dealt with were installed on high-rise buildings and allowed substantial amounts of water to penetrate the joints in the system. Substantial damage was done to the framing and sheathing, and copious amounts of mold were found throughout the cavity between the exterior gypsum sheathing and the interior sheetrock.

Dealing with Transition Issues: The Value of Counsel

Our firm has a lot of experience dealing with what are known as "transition" issues. This relates to the process by which control over the common elements of a condominium is turned over from the developer to the unit owners. Not all states have the transition process, and requirements vary among those that do. For example, in New Jersey, the transition process is triggered by the sale of at least 75 percent of the units in the condominium. The transition process is very complex and results in much litigation because of the divergent interests of the developer on the one hand and consumers represented by their condominium association on the other hand. The developer has typically sold out—or is close to selling out—the project and has already gotten its money out of the development. It is looking to wind down, get its bonds released, and get out of the project quickly so it can move on to the next deal. The condominium association has usually just hired its own independent engineers and property manager to do an objective evaluation of the condition of the common elements and finances of the association.

Most of the time, the unit owners who now control the board of directors of the association have no experience with construction and are just getting familiar with the construction issues at the same time the developer is heading for the door. Normally, by the time the association finds out about design and construction deficiencies and defects, the developer is long gone and there are no more units left to be sold. The developer is typically a shell company set up as a corporation or limited liability company that has no assets and no real interest in doing repairs. Sometimes the developer entity is owned by a reputable builder that may have some interest in protecting its name and reputation, and therefore may be willing to at least discuss doing some repairs. Unfortunately, for practical business reasons addressed below, the developer is almost never willing to spend what is required to fix what is wrong to the satisfaction of the association. That means the association has no choice other than to litigate or special assess the unit owners hundreds of thousands or millions of dollars to effectuate repairs. We counsel associations on how to deal with these issues.

The association is usually in a big hole. It has neither the money nor the experienced leadership to handle the complex process of hiring qualified experts to advise it on the technical construction and financial issues it is confronted with. Counsel therefore has to assist the board of directors of the association by recommending competent engineers and/or architects who can give the association reliable, unbiased reports on the condition of the common elements. Counsel also has to recommend to the association accounting experts who can advise the association on various financial issues including but not limited to how much money the developer may owe the association for (a) unpaid monthly maintenance fees the developer should have paid for developer-owned units prior to their sale or (b) benefits derived by the developer-owned units from common elements the association had to pay to maintain.

Selection of Experts

A lot of careful thought has to be given to what claims are worth fighting about, because the association will likely have scarce resources available to it. Much of the value counsel brings to this process is its experience and judgment, and its relationships with experts who are particularly skilled in disciplines that are responsive to the needs of the association.

Counsel also adds value to the process by making sure the association does not take unrealistic positions regarding the claims it makes. This means keeping client expectations within line of what we think we can probably get through litigation or mediation of each claim. It helps to have considerable experience when you are trying to give advice to a client about what their reasonable expectations should be regarding the amount the association can realistically expect to recover from the litigation. This is one of counsel's most important responsibilities, because the decision of the client regarding how much to spend on the case and what to settle the case for are dependent upon counsel's advice. Myriad factors need to be considered in making this critically important evaluation. Are there statute of limitations or statute of repose issues? If so, how much of a concern are they? How good are the plaintiff's experts? How strong are the plaintiff's proofs on liability and damages, especially consequential damages (as discussed in detail below)? Can the proofs be understood by the jury? Do the defendants have any chance of dismissing all or part of the plaintiff's case on pre-trial motions? How long will trial take, and does the client have the money to fund the case through trial? If not, is counsel willing to advance the costs and expert fees and take a piece of the recovery? Do the defendants have any assets? How much insurance coverage is there? What coverage issues are there, and how can the plaintiff overcome them?

Counsel has to see and anticipate issues well in advance to prevent a client from spending huge sums of money taking discovery and then having claims unexpectedly dismissed before trial. For example, suppose you have a construction defect case involving a defective exterior cladding that was negligently applied by the applicator. The building was substantially completed in 1993 and the product was applied in 1992. The developer did not immediately sell all of the units. Instead, the developer rented out the majority of the units in the building. Eventually, in 2005, twelve years after substantial completion of the building, the developer sold off the last units needed to trigger transition of control of the association's board of directors. The unit owner-controlled board hired experts who discovered that severe water penetration through the defective cladding caused mold and other massive damage to the structural framing of the building. You sue the developer, all subcontractors, and design professionals involved in selecting the material, and the manufacturer and distributor of the material.

This case is a minefield of complex issues. Counsel will have to evaluate how the statutes of repose and limitations will impact the handling of this case. If the defendants can prove the developer-controlled board of directors of the association knew about the water penetration and failed to file suit beyond the statute of limitations, it is possible the unit owner-controlled association could have all of its claims against all subcontractors, design professionals, the manufacturer, and distributor dismissed. This would leave the association with a remedy only against the developer and general contractor on the theory that during the period when they were in control of the board of directors of the association, they had a fiduciary duty to bring these claims against all possible defendants. Having failed to do so, they should be stopped from asserting the statutes as a defense.

Counsel has to be sensitive to these kinds of issues, because community associations and other kinds of clients are simply not going to be sophisticated enough to understand they even exist, let alone know how to evaluate and deal with how they impact the association's reasonable expectations for the value of the case, or how much the association should be willing to invest in attorneys' fees, expert fees, and other costs in order to get a fair return on their money.

Communicating with the Client

One other way counsel can add value for the client is by helping the association's board of directors inform the unit owners about the litigation. This takes a lot of pressure off the board and allows the community to see and hear from the attorneys who are handling the case a first-hand report on the progress of the litigation. Counsel must be mindful that these open public meetings may not be covered by the attorney/client privilege. Therefore, considerable care has to be given to making sure nothing is disclosed in these meetings that could be construed as waiver of the privilege. Handled skillfully, a public meeting at which counsel and/or experts give a top-quality presentation about the case and answer questions can be an opportunity for the board to galvanize public support for the case. This can be particularly important where a special assessment is being contemplated to pay for the litigation.

Financial Implications of Construction Law: Insurance

We have all heard Cuba Gooding in the movie *Jerry Maguire* screaming the phrase, "Show me the money!" The developer is often a shell with no assets. It may even be bankrupt. The subcontractors often are unknown or, if known, may operate out of the back of a pick-up truck and have no tangible assets. The design professionals typically have no substantial assets either. The manufacturer of defective building products may be out of business, bankrupt, or have liens against its assets in favor of a commercial lender. Faced with such scenarios, plaintiffs are often in despair about how they are going to recover millions of dollars in damages for defective materials or deficient workmanship. They are especially concerned about laying out enormous sums for legal and expert fees and costs with no idea where the money is going to come from to pay their damages. One of the most important things counsel does is show the client where the money is going to come from. The answer is insurance.

Builders, their subcontractors, and design professionals usually have commercial general liability insurance policies that pay for property damage when there is an occurrence as defined under the policy. The property damage cannot be damage to the workmanship or materials provided by the particular contractor who did the work in question. Thus, if the roofer was negligent in applying terra cotta roof tiles over sheathing applied by the framing contractor and the tiles get damaged, that is not property damage covered under the roofer's commercial general liability, because it is damage to the materials and workmanship of the roofer.

However, if the deficient installation of the roof tiles allowed water to penetrate the roof tiles and damage the sheathing underneath that was installed by someone else (i.e., the framing contractor), that is property damage covered under the roofer's commercial general liability. This is known as "consequential property damage." It does not matter whether the contractor is still in business, bankrupt, a deadbeat, or has run off to Fiji and cannot be found. As long as there was a commercial general liability in place when the deficient work was done and the insurance company is still in business, there is coverage—and a pocket to pay for the damages. In fact, if the contractor cannot be found, the plaintiff can often get the court to allow the plaintiff to serve the carrier for the absent defendant with the complaint and force the carrier to defend, and ultimately pay damages for, the absent contractor. Obviously, this is a gross oversimplification of an extremely complex analysis and the exclusions and other language of the policies will have to be carefully evaluated by experienced counsel before any conclusions about coverage can be reached. Nevertheless, counsel should be mindful of the availability of insurance coverage as an avenue of recourse for clients who have been badly hurt by deficient workmanship and/or defective building products.

The Value of Contingency

Handling a major construction litigation case takes three to five years in most state courts, sometimes longer. A complex case with fifty or more defendants can cost upwards of $1 million in legal fees plus hundreds of thousands more in expert fees and costs. Most clients cannot afford to finance such litigation. As a result, in appropriate cases, we will handle large, complex, multimillion-dollar cases on a full or partial contingency. This decision is made on a case-by-case basis depending upon many considerations such as (a) the facts; (b) the projected legal issues; (c) the insurance coverage issues we can anticipate based upon our experience; (d) what experts we will need and approximately how much they will cost; (e) how long it will take and how much we think the case will cost to get to trial; and (f) what we think the range of anticipated recovery may be. This enables the association to minimize the need to upset the members of the association with enormous special assessments during the pendency of the litigation. It also enables counsel to assign all of the attorneys it needs to the case without worrying about whether the client can afford it.

We are often able to assign two or more partners and multiple associates to complex construction litigation cases. Therefore, we can move the cases ahead much more quickly than would ordinarily be the case if the client had to pay hourly for counsel's time. The client does not care how many attorneys work the case, because the client is not paying hourly fees. In fact, the client is most often quite impressed with the effort counsel is making. This tends to make for a very close working relationship between counsel and client, because the client sees counsel is investing an enormous amount of resources into the case. In effect, the lawyer has become a true partner of the clients in the case.

Counsel gets the added benefit of putting the most talented attorneys on those aspects of the case they are best suited to handle, without regard to what that will cost the client. This enables counsel to work a very complex case with great efficiency, skill, and speed. In the end, both the client and the plaintiff's lawyers benefit from a tightly handled case that realizes a great result—and in less time than would be the case if counsel was constrained by concerns about the ability of the client to pay the legal bills every month.

Case Intake: Applicable Statutes and Insurance

When evaluating a new case, we look at many factors. One of the first considerations is applicable statutes of limitation or repose. The statute of limitations is a statutory limit on when a claim can be brought. It is an equitable statute. The case law allows the statute to be extended if a reasonable person could not, through the exercise of reasonable diligence, have known about the defect. This is known as the "discovery rule."

A statute of repose is different in that, rather than allowing a claim to be brought within a specified number of years that can be extended by the discovery rule, it simply says that within X number of years after substantial completion of work, no claims exist, ever. For example, in New Jersey, the statute of limitations on negligence claims resulting in property loss is six years, which is then extended by the discovery rule. The statute of repose in New Jersey, and in many other states, is absolute, even if the builder, subcontractor, or design professional intentionally and fraudulently concealed known defects. There is no extension of this statute under any circumstances, and the discovery rule is inapplicable. Thus, at the case intake stage, it is absolutely imperative that counsel be familiar with these statutes and understand how they are going to deal with arguments based upon those statutes.

Another important factor in evaluating a new case is available insurance policies for the general contractor, subcontractors, and design professionals. We recommend that clients who have reason to believe there is no insurance hold off on spending substantial money on expert and legal fees until the availability of insurance is determined in discovery. Many states have court rules allowing quick discovery of insurance information, a simple process that can provide critically needed information early in a case. If the case is being taken on a contingency, it is a good idea to have an agreement in advance of filing suit that counsel will hold off aggressively litigating until counsel gets the insurance policies in discovery and can make certain there are no obvious exclusions that destroy coverage. For example, if you are litigating over damages caused by EIFS, you will want to know as quickly as possible if the developer and the EIFS applicator have insurance policies that contain EIFS exclusions. If it turns out there is no insurance, the case can be dismissed before the association spends substantial money on litigation that is going to be fruitless in generating a recovery.

The Importance of Experts

Another important consideration is whether the association is going to be willing to hire and pay for the qualified experts necessary to win the case. You can have a great case, but if you have the wrong expert, it can quickly turn into a disaster. The best way to avoid this problem is for counsel to give the client a thorough and honest estimate of what the case will cost and what the timing of the case will likely be.

Many lawyers file suit based upon transition engineering reports. These reports are often perfunctory and were done just to give the association a general idea of what issues might be of concern. Many explicitly state that no invasive testing was done and that additional investigations are needed.

We like to get as much engineering work done as possible before the complaint is filed so we have a good understanding of the strengths of the case. For that reason, we prefer to hire experts we will rely upon at trial to do a thorough investigation of whatever defects have been identified or are suspected. These investigations involve extensive test cuts and as many photos as are needed to properly document conditions. They can take weeks to complete and can result in extensive reports that cost tens of thousands of dollars or more. Counsel has to use good judgment on a case-by-case basis to guide the client as to what level of engineering work is needed or appropriate before the complaint is filed. Generally, the more complex the issues, the more detailed the investigation will have to be. For example, if the case involves roof claims on forty-five buildings, a decision has to be made as to whether you are going to have your experts test all forty-five buildings or try to test just a representative sample. If you have issues relating to deficient installation of fire suppression systems because the wrong antifreeze was put into the CPVC pipes, causing damage from environmental stress cracks, a careful analysis will need to be made as to what type of experts you need. Depending upon your proof, if you do not have burst pipes yet, you may need to hire an expert who is a chemist and can testify about environmental stress cracks and how they can or will cause consequential damage in the near future. The list of possible examples is limitless. The point is that close consultation between the clients, counsel, and experts is obviously needed in making this judgment.

Use of Experts in Construction Litigation

Construction litigation requires the use of experts in many contexts. For example, a typical project starts with the design phase in which civil engineers, geotechnical engineers, hydrologists, environmental engineers, surveyors, and architects are involved in shaping the contours and features of the project design. The design phase is followed by the construction phase in which numerous design professionals are involved. These include, among others, architects, structural engineers, mechanical engineers, and electrical engineers. All of these design professionals must coordinate their activities with those of the general contractor and subcontractors.

Design professionals also have to be careful to make sure materials specified for the project are fit for their intended purposes and will not cause or exacerbate construction deficiencies. Many construction materials become code-compliant as alternate materials under the building code, and their manufacturer's installation specifications become the standard for code compliance. Communication with the manufacturer or distributor becomes very important in determining whether materials can be used on a particular project.

Assuming litigation ensues after design or construction defects are found, experts will almost certainly be required to render reports and testify at depositions and at trial. Counsel will also need to consult with experts during various stages of the case. Typical issues that may require use of expert testimony or require counsel to consult with experts include, among many others:

- Whether work was performed deficiently
- Whether materials were defective
- Whether certain work was within the scope of work of a particular contract
- Whether the work of a contractor conformed to the scope of work in the contract
- Whether the design or workmanship was deficient or the materials were defective, or some combination of the three
- The applicable standard of care governing that work
- Whether there were delays that are actionable
- The cause of the deficiency or defect
- The damages flowing from delays or deficient workmanship or materials
- What needs to be done to repair the defective materials or workmanship

Selecting the Consulting or Litigation Expert

There are two classes of experts: consulting experts who will not be testifying at trial and trial experts whose testimony you do intend to rely upon. All documents and oral communications exchanged between counsel and the consulting expert are generally privileged, except in rare instances where the opposing party can show a hardship that justifies allowing the consulting expert's files to be discovered or that allows the consulting expert to be deposed. That would only happen in a situation where the opposing party could show there was no other way he or she could get anyone else to offer an opinion on the same issue (for example, because the evidence was destroyed). In New Jersey, discovery is allowed into the documents and oral communications between the testifying expert and counsel to the extent, in formulating his or her opinion, that the expert relied upon any information exchanged or communicated between them. This distinction is important, because counsel's communications with consulting experts—even those who led to the development of opinions harmful to the case—are not discoverable. This gives counsel the freedom to be aggressive in speaking with prospective experts so counsel can find the expert best suited to the case.

There are many factors to consider in hiring an expert. The nature and value of the claim, the factual background of the matter from which the claim arises, and the scope and timing of the work must be considered.

The credentials and experience of the expert must also be considered since the fees charged by the expert usually are a direct reflection of those factors. You must also decide whether the expert is going to be a consulting expert or an expert who will be testifying at trial. This is a particularly important consideration given the privilege that attaches to communications between the attorney and the consulting expert.

Some or all construction and design claims may involve extremely complex factual or technical issues. The entire case may rise or fall on the quality of the experts selected to handle the claims in issue.

Extraordinary care must therefore be given to the entire process of finding and hiring consulting and testifying experts. The initial expert retained by the client may be an architect or engineer with broad experience in areas of design or construction. As counsel's investigation continues with the assistance of the consulting expert, it may be necessary to bring in additional experts who have specific expertise in particular areas. The amount in controversy will often have a major impact on this process. If the case is a construction and design defect case involving water infiltration from roofs, windows, and siding, and claims relating to improper design of a sea wall, and deficient construction of the fire suppression systems in a large condominium, and the amount in controversy is $5 million, it may make sense to engage an expert with particular expertise in handling just the roofs, windows, and siding, a different expert with particular expertise in sea wall design and construction issues, and another expert with particular expertise in design and construction of the type of fire suppression system involved in the case. Counsel may also wish to hire yet another expert with substantial expertise in cost estimating to compile the damages report.

As a general matter, it is advisable to make sure your experts have experience testifying in depositions and at trial. That is a very different experience for most experts. No matter how well educated and experienced an expert may be, it is not unusual for an expert who has never testified before to look tentative, become defensive or combative, or otherwise have difficulty under cross-examination by skilled counsel. If the claim is worth bringing and is substantial enough for the client to incur the expense of hiring an expert, it is worth the time and trouble to make sure that, in the interview process, due consideration is given to making sure the prospective expert has testimonial experience. These considerations all need to be thoroughly reviewed with the client before decisions are made.

Avoiding Trouble: Spoliation of Evidence

When engineering work is being done, careful consideration must be given to the doctrine of spoliation of evidence. This is a doctrine that says that before any repair work or invasive testing is done that substantially affects the original condition of the property in question, fair notice must be given to any person who may have an interest in that work. Failure to give fair notice in advance of the work can mean the plaintiff doing the work can have its proofs barred at trial.

The point of the spoliation of evidence doctrine is to make sure contractors, material suppliers, design professionals, and anyone else who may eventually be sued for design- or construction-related claims has adequate notice and a fair opportunity to view the existing "as-built" condition of the property before any repairs or major invasive testing are done. The theory is that if such a person is deprived of the opportunity to inspect the as-built condition of the property, the person is essentially deprived of the ability to mount an effective defense or to make claims against third parties it believes are responsible for the loss. Notice should be given if the property is about to be repaired. It may also be necessary if extensive invasive testing is going to be done.

For example, if you have a client who is going to demolish extensive portions of one side of a building to see what the condition of the sheathing or framing is, or to observe and document mold conditions, notice should be given. If your expert is just going to do some small test cuts of the exterior cladding, notice probably does not have to be given since you are not affecting the ability of the defense experts to either reopen the test cuts done by your expert or do test cuts of their own.

Giving notice is a judgment call for counsel. We recommend that counsel err on the side of caution and give as much notice to as many people as possible. For example, we do a lot of cases involving deficient installation of building envelopes. We often have no idea who the manufacturer of the defective exterior cladding is. Before we allow our clients to tear off the cladding and fix their buildings, we give notice to every product manufacturer and distributor we can think of. We also ask the developer or general contractor to give written notice to all contractors and design professionals who worked on the project. By being as expansive as possible, you limit your exposure to the spoliation of evidence defense.

Who Do You Sue?

Generally, in determining who to sue and what claims to make, we are guided by our client's knowledge of the history of the project, our experts, and our experience. We also review whatever documents are available and make reasoned judgments as to who to join in the original complaint. It is especially helpful to have the original as-built plans, the records of the construction office of the local municipality, and the records of any other local, state, and federal agencies that may have played any role in approving the design of the development or project in question. We understand we are going to have to amend the complaint—probably several times—until we have all parties joined who belong in the case.

We start with the developer and general contractor and then add any subcontractors whose identities are known to us and who we have a good faith reason to believe have liability for the defects in question. If there are any defective products, we try to determine who the manufacturer and distributor are so we can join them to the suit. We frequently ask the developer or general contractor for a list of responsible subcontractors. They are usually willing to provide this information, since they are typically looking to be indemnified by these subcontractors and design professionals anyway under the terms of the contracts between them or under state law.

Once the complaint is filed, we serve extensive requests for documents relating to, among other things, the approvals, design, construction, and inspection of the project, insurance coverage, and any deficiencies discovered or complaints by any unit owners, homeowners, or other buyers or residents. We also serve written interrogatories on all parties to elicit important information such as (a) who did what in connection with the project; (b) who supplied the materials in question; (c) who inspected the work; (d) what changes were made in the work; (e) what problems arose during the project; (f) what payments were made for the work or materials in question; and (g) what disputes arose during or after the project. After several months of discovery, we can usually identify most of the important players. We do additional amendments of the complaint as we go forward and identify other responsible parties such as subcontractors of subcontractors and repair contractors who may have come in after the job was well underway to fix someone else's deficient work and thereby exacerbated already deficient work.

Plan Ahead: Have a Strategy

Joining parties in a complex construction litigation case involving millions of dollars of damages is not to be done lightly. If counsel is not careful, it is easy to wind up with fifty or more parties, each of whom has to be served with every pleading, motion, letter, discovery request, expert report, and everything else that is sent out in the litigation. This can cost tens—sometimes hundreds—of thousands of dollars each year the case continues.

If a party is going to be joined, careful thought must be given to what it will cost to add that party and what recovery is likely to be obtained. We have seen complex cases where a party is joined because of a relatively small claim. That party then files a third-party complaint and brings in two other parties seeking contribution or indemnification. The three of them get involved in a series of disputes about who did what on the project. Each party serves extensive discovery requests on the plaintiff and on each other. Experts have to be hired for each party, and site inspections have to be arranged. By the time everyone gets done with all of that work, the amount to be recovered has been dwarfed by the amount of attorneys' and expert fees and costs expended. This causes aggravation and delay for all involved, and in the end, no one is happy with the outcome. Thus, even if the plaintiff has a perfectly valid claim, careful consideration has to be given as to whether it is economically sensible to assert that claim.

Generally, experienced counsel knows you follow the money in a construction litigation case. Common sense dictates that you assert and spend most of your time on the most valuable claims. You must make sure you have a good understanding of insurance coverage issues as well as the best experts you can find. Care has to be taken to plead your claims in such a way as to make it as likely as possible that the insurance carriers will refrain from disclaiming insurance coverage. For example, you will want to try to frame your pleadings to describe negligent rather than intentional conduct. Typically, commercial general liability policies exclude claims for intentional conduct but cover negligent conduct. In a similar vein, counsel will want to carefully consider whether he or she wants to plead fraud claims, because they involve proof of intentional conduct that is beyond the scope of commercial general liability policies.

Once the initial investigation and preparation of pleadings is complete, counsel has to assimilate the documents and interrogatory answers and then get ready for depositions. This is the most important part of discovery, because you are getting answers directly from the witness without counsel coloring the answer, as typically occurs when interrogatories are answered by counsel and then affirmed by the client.

Depositions

Depositions are a critically important part of the plaintiff's strategy. They enable the plaintiff to bring out facts that trigger insurance coverage. They also enable the plaintiff to obtain testimony that, among other things, (a) authenticates important documents; (b) lays foundations for other testimony; (c) provides fodder for impeachment of witnesses at trial; (d) gives the plaintiff the opportunity to create a record that can be used to defeat summary judgment motions at the close of discovery and to support the opinions of its expert; and (e) sustain its burden of proof at trial. We often spend months preparing for depositions of important witnesses such as project managers, design professionals, and product manufacturers, since those depositions can be of seminal importance. A great deal of thought, preparation, experience, and skill is required in planning and executing your deposition strategy, because those depositions—properly done—set up the entire case.

We spend substantial amounts of time and money doing intensive investigations of non-parties and their documents. Disgruntled former employees, competitors, and municipal officials can be a fountain of useful information. In addition, representatives of manufacturers of materials that were used on the project often do site inspections that result in interesting meetings, letters, e-mail, and other documents that can be extremely useful at trial. Counsel should do plenty of thinking about discovery strategy, because it can turn a valuable case into a goldmine if properly conceived and executed.

We are amazed at the lack of knowledge of many project managers and superintendents we depose. These project managers and their subordinates are entrusted with supervising the work of hundreds of people costing tens of millions of dollars. We typically prepare by learning every nuance of the details and specifications governing the issues in the case. We spend quite a bit of time mastering those details and specifications, because we have learned over time that the project managers and their subordinate superintendents typically do not have much understanding of those details and specifications. Devastating testimony can be developed by going through those details and specifications with the project manager and his or her subordinates. The obvious goal is to elicit testimony from them that enables counsel to put on testimony at trial raising the question of how you can possibly evaluate the acceptability of the work of the subcontractors in a competent way if you do not understand what they were supposed to do.

One final example is worth describing. You never know where your next bombshell is going to come from. That is why we try to talk to as many people who worked on the project as possible. We spend a great deal of time and money tracking down missing subcontractors, especially those who are out of business and may have a good story to tell. Disgruntled former employees and subcontractors are a veritable fountain of valuable information about what transpired on a job. We have taken depositions where disaffected subcontractors who were not paid a few thousand dollars and were thrown off the job come back years later at depositions and do enormous harm to the developer or general contractor. For example, in one particular instance, the now-defunct subcontractor responsible for applying the roofs to a series of condominium buildings was upset because he did not get paid his last $20,000 of a contract worth more than $1 million. When asked why he did not follow the manufacturer's installation specifications for installation of ice and water shield and why he failed to use ventilated, pressure-treated stirrups to attach the concrete roof tiles, he threw the project manager under the proverbial bus with great enthusiasm. He testified that he pointed out the details and specifications of the roof tile manufacturer to the project manager, but the project manager told him to ignore them because the builder did not want to spend the money for these details. He even went so far as to get his ex-wife, who held the defunct company's records, to give us documents he had that we used to help support his story. The carriers for the roofer and the general contractor eventually threw in their full policies to settle the claims.

It takes years of experience to be able to effectively handle a complex construction litigation case involving claims arising from design defects and/or construction deficiencies. These cases are an interesting challenge and a great deal of fun to handle. There is nothing quite like being the plaintiff representing completely innocent, totally victimized clients in a complex case involving tens of millions of dollars in claims and appearing at depositions, court hearings, and trial with fifty or more lawyers arrayed against you. The fact that your clients had nothing to do with creating any of the construction deficiencies and design defects in question gives you a powerful moral and legal advantage over the defendants. Careful thought, planning, preparation, use of the right experts, lots of hard work, and hopefully some good luck can get you a result that is emotionally fulfilling and financially rewarding for both your client and your firm.

Donald B. Brenner is a shareholder in the business litigation group at Stark & Stark and chair of the firm's construction litigation group. He has broad experience in handling complex commercial litigation with an emphasis on construction litigation and litigation on behalf of community associations. He has been extensively interviewed on television about exterior insulation and finish systems and construction litigation cases, including on WABC-TV in New York and the "CBS Evening News" with Dan Rather, and he has been quoted in The New York Times. *Mr. Brenner is the co-founder of www.njeifs.com and received intensive training in exterior insulation and finish systems at the Exterior Design Institute. He also lectures extensively on exterior insulation and finish systems and other construction litigation issues.*

Mr. Brenner earned his B.A., magna cum laude, from the State University of New York at Albany and his J.D. from Rutgers University School of Law, where he was a notes and comments editor for the Rutgers Law Journal. *He is admitted to the bars of New Jersey and Pennsylvania, and he is a member of the American Bar Association and the Association of Trial Lawyers of America. He is also a contributing editor to the* Commercial Law Journal *and a former contributing editor to* Community Trends, *the magazine of the Community Associations Institute.*

Understanding the Project and the Client's Goals

Michael Magee, Esq.

Partner

Pietragallo, Bosick & Gordon LLP

The Construction Lawyer's Role

As a construction lawyer, I provide a broad range of legal services to clients throughout various phases of their projects, from inception to wrap-up. These services include entity formation and structuring, contract negotiation and preparation, risk management, and litigation or alternative dispute resolution.

With respect to adding value for clients, it is best for a construction lawyer to be involved from the project's inception. When this is the case, it allows counsel's principal focus to be on structuring the project in a way that best protects the client. Of principal concern at this stage is managing project risk including potential claims relating to personal injury and environmental and property damage. These risks are a major cost factor to the construction industry, the effective management of which will lead to increased profitability.

When we become involved at the dispute stage, my focus is on achieving the best result possible for the client in the most cost-efficient manner. This is achieved through aggressive and efficient case management and utilization of effective cost management controls.

Ultimately, my success is defined by my clients' successes, not only on a project basis, but with respect to meeting their overall business goals and ensuring their satisfaction with the ultimate outcome. In the role of construction lawyers, success is only possible for those attorneys who are able to maintain their edge in this field. It is imperative to stay on top of current and emerging trends and information. I do this through day-to-day discussions with people in the industry and by reviewing and analyzing current case law from around the country. I also attend industry-related seminars and participate in industry-related professional groups.

Preparing for the First Client Meeting

When planning to meet a new client, the information my team and I need depends somewhat on the timing of the meeting. If the meeting is taking place at or near project inception, I want to know details about the scope of the project, its timetable, the client's role in the project, and the overall

UNDERSTANDING THE PROJECT AND THE CLIENT'S GOALS

project value, as well as its value to the client. I also would find out the number and identity of general contractors and subcontractors involved, including their timetables and bid requirements.

If, however, the initial meeting is taking place once litigation has been or is about to be commenced, the type of information I would need changes slightly. If I were representing the defendant, I would need to obtain the filing date of any claims, the nature of the claims asserted, the identity and role of the plaintiff(s), the name and location of the court where the case is filed, and the award or remedies sought. Another component I would seek to determine is the availability of insurance coverage or indemnity. If I were representing the plaintiff, I would need to determine the potential theories of recovery, the identity and role of potential parties, remedies to be sought, the nature and value of potential claims, and the financial viability of potential defendants.

Obtaining this information is critical, because it provides an overall idea of project scope and gives me an understanding of the players involved and the client's roles in the project. By the time I meet with the client, this information will enable me to discuss the project knowledgeably and to prepare a list of potential issues and concerns that must be addressed.

Information to Obtain at the First Meeting

Information that must be obtained during the initial meeting depends upon the phase of the project at which the attorney is joining the proceedings. For example, if the lawyer is involved from the inception of the project, critical information to obtain includes an understanding of the client's level of interest in the project, including their comfort level in performing the contract. Other concerns are bonding, schedule, labor, lien, indemnity, and insurance issues as well as the client's relationship(s) with other contractors on the project.

If counsel is joining the proceedings once a dispute has already reached litigation, some important facts to obtain from the client include the project timeline, factual background, litigation goals, key client contacts (i.e., the employee or contact who is the most knowledgeable about the project), and the nature and components of amounts claimed or to be claimed and/or the remedies sought. You will also want to understand, to the best of your ability, the strengths and weaknesses of the claims and the state of all documentation and contractual provisions impacting the claims.

On the business side, this information helps navigate contract negotiation and preparation. On the litigation side, it will assist in outlining litigation strategy and discussing the viability of the claims and defenses.

In terms of developing a customized strategy, the information that comes out of the initial client meeting will help construct the strategy that is best suited for the issues facing the client. With respect to litigation, the information that is initially gathered will help in determining the overall strengths and weaknesses of the case. It will also aid in determining the direction in which the case should proceed, whether it be mediation, arbitration, settlement, or traditional litigation. Finally, some components of this critical information will also help in identifying and setting achievable goals based on the client's desires and counsel's assessment of the realistic outcomes of the situation.

Determining Strategy: General Questions to Ask Clients

The general approach for a lawyer representing a client involved in a construction dispute is to determine the most cost-efficient means of achieving the client's goals. In most cases, this will mean achieving a satisfactory resolution of the dispute in the shortest amount of time. By nature, such a speedy and cost-effective resolution would involve minimal discovery, particularly with respect to the document production aspects of construction litigation.

However, each case also needs a specific and customized approach based on the client's goals, resources, and unique circumstances. In general, the attorney should determine what the client ultimately wants to achieve in the dispute resolution process. You will also want to determine the client's comfort level with the evidence (both documentary and testimonial) and how it supports their position. Generally, this can be achieved simply by having them relate their understanding of the facts involved in the dispute to you. It is also important to determine how committed management and/or ownership is to pursuing the claim, since a lack of commitment often adversely affects the overall success achieved in the dispute. Is it a matter to which they are committed to litigating through to jury verdict if necessary, or is it a matter they wish to settle?

The motivation behind asking these specific questions is the need to establish that the client has reasonable expectations regarding the outcome of the litigation. Further, the answers to these questions help ensure that counsel gains a true understanding of the nature of the dispute.

Essential Documentation for Clients to Provide

Depending on the type of dispute, the following documents are essential in understanding the dispute: bid documents, including internal support documentation; contract documents, including all extra work orders; all written communications pertaining to the project, including e-mail, instant messages, fax and mail logs, employee project diaries/notes/calendars, daily job records, and scheduling documentation; all applicable insurance documentation; bonds; payroll and staffing records; project photographs and images; management organizational charts; drawings; documentation of timetables and/or critical paths; and billings and payment records.

In addition to all of this documentation, the construction lawyer must have access to critical software such as budgeting, critical path methods, billing, and project management. All of this material is important because it paints an overall picture of the project in addition to providing the critical factual details needed to litigate the dispute. While this documentation will not always necessarily help the case itself, it will always help build counsel's understanding of a case. Once counsel has a thorough understanding of the case, he or she can better advise the client on the strengths and weaknesses of the case, avenues that should be pursued or avoided, and the likelihood of prevailing or, in contrast, the need for early resolution.

Establishing a Positive Client Relationship in a Construction Matter

In any legal matter, it is important for counsel to establish a positive relationship with the client. The first step to doing so is to make sure to understand the project itself as well as the client. Make sure the client understands that he or she is the client and, while counsel will provide the best legal advice possible, it is ultimately the client who makes the decisions and controls the litigation. Assure all clients that there will be no surprises: all information will be relayed to them in a timely and thorough manner.

Another way to establish solid relationships with clients is to form a strong team that consists of the client, lawyer, and other professional advisors. Early on, I work with the client to establish clear client expectations. At this phase, I also demonstrate an awareness of the need for cost-efficiency. Ultimately, every client is treated as if he or she is my only client, which goes a great way toward establishing a positive rapport based on trust.

This strategy is successful because construction clients need to be intimately involved in their legal disputes, and for the most part, well-informed clients who are actively involved in the litigation process tend to make better decisions.

Clients' Goals in Construction Matters

Learning your clients' true goals is essential to developing an appropriate strategy. In litigation, counsel should simply ask clients what they think their ultimate goals are, as well as what they think it would take to reach them and, most importantly, what they are willing to do to get there.

In instances when you find a client's goals to be unrealistic, the lawyer simply has to tell them the truth. Once you tell the client their goal is unrealistic, explain why and offer what you believe to be a realistic goal. Most construction clients expect you to be brutally honest with them.

It is essential to understand a client's true motives and goals when developing a strategy for a given construction matter. The dangers of failing to do so include reaching an adverse outcome and finding yourself with a dissatisfied client.

Developing the Theory of the Case

I develop the theory of a case by meeting with the client, reviewing and discussing documentation with key players, and then ultimately applying the law to this information.

In developing the theory of the case, you must also take into account the non-legal ramifications of the issues facing the client. In most instances, economic ramifications to the client have a very strong influence not only in conduct of litigation but also in its ultimate resolution. These broader economic considerations often drive the ultimate resolution of construction disputes.

Ultimately, successful representation of a construction client boils down to counsel's understanding of the project and the client's goals. Once these are understood, developing and implementing a successful strategy is a matter of diligence in the gathering and analyzing of the information available to counsel.

Michael Magee is a partner in the law firm of Pietragallo, Bosick & Gordon LLP and is a member of its construction services practice group. Mr. Magee has broad and extensive experience in the areas of commercial litigation and general business matters relating to the construction industry. For more than fifteen years, he has represented contractors, developers, engineers, architects, and owners on a local and national basis in various state and federal courts, including the U.S. Court of Federal Claims, concerning a wide variety of construction-related issues. He has previously served as national litigation coordinating counsel for several construction and engineering firms and has substantial experience in the management of large-scale, multi-jurisdictional litigation.

Mr. Magee received his law degree from Duquesne University School of Law and his undergraduate degree from the University of Pittsburgh. He is admitted to practice in the state courts of Pennsylvania, West Virginia, and Texas, as well as the U.S. Court of Appeals for the Third Circuit and the U.S. Court of Federal Claims. Prior to attending law school, Mr. Magee served as assistant executive director for U.S. Senator Arlen Specter.

Appendices

CONTENTS

APPENDIX A

CONTRACT FOR ENGINEERING, PROCUREMENT, AND CONSTRUCTION SERVICES

THIS CONTRACT FOR ENGINEERING, PROCUREMENT AND CONSTRUCTION SERVICES (the "Agreement") has been made and entered into as of the ____ day of _____, 20____ (the "Effective Date"), by and between _____, a _____ corporation ("Owner"), and _____, a _____corporation ("Contractor").

RECITALS

WHEREAS, Owner desires to have constructed a new facility consisting of _____ and Associated Equipment (as defined in Exhibit A hereto) and Contractor supplied BOP (as defined in Exhibit A) (the "Plant") constructed on Owner's site located in _____ and described on Annex 1 (the "Site");

WHEREAS, Contractor has provided a technical proposal and not to exceed price for providing detailed engineering, procurement, fabrication and delivery of the balance of plant materials and equipment (collectively, the "BOP") and detailed engineering, construction, installation and start-up and commissioning services with respect to the entirety of the Plant; and

WHEREAS, Owner desires to engage Contractor as its general contractor to provide the Plant to Owner, complete and ready to operate, and Contractor is willing to do so in accordance with and subject to the terms, provisions, and conditions hereinafter set forth.

NOW, THEREFORE, in consideration of these premises and the mutual covenants and obligations hereinafter set forth, and intending to be legally bound, Owner and Contractor hereby agree and obligate themselves as follows:

ARTICLE I

DEFINITIONS; SCOPE OF THE WORK

1.1.　Defined Terms. Capitalized terms used in this Agreement shall have the meanings assigned to such terms in Exhibit A to this Agreement.

1.2.　Description of the Work. (a) Contractor shall be responsible for performing all design, engineering, procurement, fabrication, construction and other services and providing all labor, supervision, materials, supplies, equipment and other goods and services (except for those goods and services which Owner has expressly agreed to provide in this Agreement or in the Exhibits hereto) which are necessary to deliver the Plant, as shown and described in the TSD (a copy of which is attached hereto as Exhibit B) and in compliance with all applicable codes, laws, rules and regulations (the "Work"), with each Unit to be Unit Mechanically Complete by the Guaranteed Unit Mechanical Completion Date, and with Final Acceptance having occurred by the Guaranteed Plant Completion Date.

(b)　Without limiting the generality of Section 1.2 (a), Contractor acknowledges having prepared and submitted to Owner the TSD after _____, and represents and warrants to Owner that it has a sufficient understanding of the Work to have established the Maximum Contract Price realizing that the Maximum Contract Price shall only be adjusted in the limited circumstances expressly provided for herein. Contractor further acknowledges that it is required to supply, and the Maximum Contract Price includes, everything necessary (except for those goods and services which Owner has expressly agreed to provide in this Agreement or in the Exhibits hereto) to deliver the Plant to Owner in a complete and operable condition and Contractor understands that it shall not be relieved of such obligation by reason of any item of services, materials or equipment having been omitted or having been incorrectly described in the TSD or elsewhere in this Agreement.

(c)　Without limiting the generality of Section 1.2(a), in connection with performing the Work, Contractor shall do all of the following, the cost of which is included in the Maximum Contract Price:

- Manage and supervise the Work so that the Work is performed in an expeditious and efficient manner, and furnish the services of all home office and field personnel necessary to maintain cost and material control, quality control, safety and environmental protection, scheduling control and monitoring of the Work (including portions of the Work assigned to subcontractors).

- Achieve Unit Mechanical Completion and Final Acceptance in accordance with the Guaranteed Unit Mechanical Completion Date and the Guaranteed Plant Completion Date, respectively.

- Furnish the services of all necessary supervisors, field engineers, foreman, skilled and unskilled labor, and other personnel necessary to complete the Work.

- Manage the dismantling and transportation of the Units and Associated Equipment both to the refurbishers and to the Site.

- Procure and supply all materials, supplies, equipment, and construction items necessary to perform the Work in accordance with this Agreement, including without limitation the TSD, except those materials to be furnished by Owner in accordance with this Agreement and the Exhibits hereto.

- Obtain all governmental approvals, licenses and permits necessary to perform and complete the Work in the name of the Contractor for the benefit of the Owner, except those permits to be obtained by Owner as specified in the TSD (and upon Final Acceptance to assign all such licenses and permits which benefit the Owner to the Owner, to the extent assignable under applicable laws), including without limitation all building permits, electrical, mechanical, and plumbing permits.

- Prepare, update and deliver to Owner a Project Schedule, other schedules, estimates and reports as reasonably requested by Owner.

- Provide precautionary measures to prevent access of unauthorized persons to the Site and theft of materials, tools and equipment from the Site.

- Comply with all of its obligations and responsibilities under this Agreement in a diligent and timely manner, in accordance with all applicable federal, state, parish, county and municipal laws, ordinances, rules, regulations and orders.

- Provide temporary hook-up and connection to Owner supplied utilities for consumption by Contractor and its subcontractors on the Site.

- Supply all outside telephone lines for permanent plant communications and for temporary use by Contractor and its subcontractors on the Site.

- Provide construction offices for Owner's technical representatives, Contractor and its subcontractors on the Site.

- Promptly receive and offload the Units following their arrival at the Site or offsite storage location following completion of refurbishment.

- Deliver the Associated Equipment directly to the Site or offsite storage location on a timely basis following their removal from _____ and provide a list of all Associated Equipment inventoried as well as a list of all Associated Equipment that cannot be refurbished and therefore will be provided by Contractor as a part of the BOP.

- Supply all chemicals, lubricants and lube oils for the Units, Associated Equipment and the BOP equipment during Start-up.

(d)   In connection with performing the Work, Owner shall do all of the following, at no cost to Contractor:

- Provide four (4) (collectively, "Units" and individually, "Unit") and Associated Equipment, which will be transported to the Site.

- Provide one (1) Step Up Transformer (a 75 MVA transformer) and high voltage switches with associated substation and BUS extensions, and all other Owner-furnished equipment and materials as specified in the TSD.

- Provide a natural gas pipeline at the Site boundary to supply the Plant.

- Obtain all environmental permits necessary to perform and complete the Work and operate the Plant.

- Supply all lubricants, lube oils and chemicals operation of the Units following completion of Start-up.

- Provide temporary utilities, electricity and water for construction.

(e)   Contractor represents and warrants to Owner that Contractor has the necessary experience and resources to perform the Work.

(f)   Contractor agrees to promptly notify Owner in writing of any errors, deficiencies, gaps, inconsistencies, questions or omissions which may exist or be discovered between (i) the TSD and any other part of this Agreement; and (ii) the engineering design prepared by Contractor and approved by Owner and any shop drawings or other construction details provided by vendors, suppliers and subcontractors, together with Contractor's recommended resolution of such matter. If Contractor and Owner agree on a resolution, such resolution shall be documented by issuance of a Change Order. If Contractor and Owner do not agree on a resolution, Owner shall advise Contractor in writing of the resolution to be implemented and Contractor shall promptly proceed to implement such resolution, it being understood that in so implementing the resolution directed by Owner, Contractor shall not be deemed to have waived any of its rights to seek a resolution of the dispute under Article XXV of this Agreement.

1.3   Changes to the Scope of the Work. Any additions, modifications, and changes to the scope of the Work shall only be made in accordance with Article VIII.

1.4   Precedence among Documents. Other than as set forth in Section 1.2(f) above, in the event of any conflict or inconsistency, whether it be direct or indirect, between this Agreement and any Exhibit or Annex attached hereto, the terms and provisions of this Agreement shall govern and control; in the event of any conflict or inconsistency, whether it be direct or indirect, between any Exhibits or Annexes attached hereto, the most recently prepared Exhibit or Annex bearing evidence of approval by both parties shall govern and control.

1.5   Access to Site. Owner shall provide Contractor and its subcontractors, materialmen, suppliers, and vendors access to the Site to perform the Work.

1.6   Parking and Storage of Materials. Owner shall designate a parking area at the Site for Contractor, and its employees, subcontractors, materialmen, laborers, suppliers and vendors to park their vehicles. Contractor, and its employees, subcontractors, materialmen, laborers, suppliers and vendors shall only park their vehicles in the area designated by Owner. Owner shall provide Contractor with a location in which to store materials and equipment delivered to the Site, which location shall be fenced, managed and keep free of debris by Contractor. Should the Site storage location provided by Owner prove to be inadequate, Owner shall reimburse Contractor, which reimbursement shall not be credited against the Maximum Contract Price, for the reasonable costs of providing an alternate storage area near the Site, including extra costs incurred by Contractor for loading, unloading, transportation and providing security at such alternate storage area.

1.7   Liability for Vehicles and Stored Materials. Owner hereby specifically disclaims any and all liability for damage, loss or theft of the vehicles parked and the materials and equipment stored at the Site. Owner hereby disclaims any and all warranties, representations or covenants as to the safety and security of such vehicles and materials and equipment, regardless of whether imposed by law, in equity, by contract or otherwise, and Contractor hereby waives any right to recover against Owner for any loss or damage to property or injury to persons arising out of such use of portions of the Site to the fullest extent allowed under applicable laws.

ARTICLE II

REPRESENTATIVES OF OWNER AND CONTRACTOR

2.1.   Owner's Representatives.

(a)   Owner may appoint up to five (5) representatives, one or more of whom may be the employees of an independent engineer, who shall be resident on a full or part time basis at Contractor's offices and fabrication facilities and at the Site during the performance of the Work. Contractor shall, at Contractor's expense, provide furnished office space, telephone and telecommunications facilities, and access to secretarial services for such representatives. Owner's representatives shall have access to the Work (and to all drawings, specifications, plans, shop samples, design, engineering and fabrication materials comprising part of the Work), at all times including, but not limited to, the right to reasonably engage in technical discussions with Contractor's engineering and other personnel performing or supervising any part of the Work, with the understanding that such discussions necessarily must be scheduled and conducted in a manner that will not interrupt or unnecessarily impede the performance of Work by the Contractor. Owner's representatives shall also be allowed to attend meetings, for information purposes only, with subcontractors, vendors, materialmen and suppliers, whether at Contractor's offices and fabrication facilities or elsewhere. Any actual instructions, rejections, and approvals to subcontractors, vendors, materialmen and suppliers that are deemed necessary during the course of such meetings will be provided by Contractor and all activities during such meetings will be in keeping with the provisions of Section 26.3. If Owner desires to issue any instructions to a subcontractor, vendor, materialmen or supplier, Owner must submit such instruction to Contractor who will not unreasonably withhold issuing the requested instruction to the proper person. Owner may replace one or more of its representatives at any time and for any reason following notice to Contractor. Nothing contained in this Section 2.1(a) shall be construed as imposing any obligation or liability on the Owner to the subcontractors, vendors, materialmen or suppliers, or relieve Contractor of its obligation to properly perform the Work on a timely basis in accordance with the terms and conditions of this Agreement.

(b)   Owner shall designate one of its representatives as its principal representative and such principal representative shall be authorized to act on Owner's behalf in all technical matters such as approval of drawings, plans and specifications, approval of subcontractors, vendors, materialmen and suppliers, the execution of Change Orders, rejection of portions of the Work, approval of Contractor's invoices and supporting materials, and similar matters not resulting, except in the case of properly executed Change Orders, in an adjustment to the Milestone Dates, the Maximum Contract Price, or any other provision of this Agreement (none of which may be modified, amended or waived except in the manner specifically provided under this Agreement). Provided that Owner's principal representative does not exceed the grant of authority contained in the preceding sentence, Contractor may at all reasonable times consult with and rely upon decisions made by such principal representative of Owner. Owner may replace its principal representative at any time and for any reason following written notice to Contractor.

2.2.   Contractor's Representative. Contractor has chosen the individuals listed on Annex 2 for the functions listed next to each such individual's name on Annex 2, all of which have been approved by Owner. Contractor shall not replace any such listed individual without Owner's prior written approval. Owner may, however, in its reasonable discretion and at any time, require Contractor to replace one or more of the individuals listed on Annex 2 with alternative employees. Contractor shall also timely appoint a project manager who shall be authorized to act on behalf of Contractor in all matters relating to the Work (the "Project Manager"), and Owner may at all reasonable times consult with and rely upon decisions made by such Project Manager. The individual appointed by Contractor as its Project Manager shall be subject to the prior approval of Owner and such individual, when appointed, shall not be removed or replaced except upon the request of or with the approval of Owner, which approval will not be unreasonably withheld or delayed. Owner may, in its reasonable discretion and at any time, require Contractor to appoint an alternative Project Manager.

ARTICLE III

SUBCONTRACTORS AND VENDORS

3.1.   Subcontractor and Vendor Selection. (a) Owner acknowledges that Contractor shall subcontract all or a portion of the engineering and construction portions of the Work (each a "Subcontract"). Attached hereto as Exhibit D is a list of subcontractors and vendors which have been pre-approved by Owner for portions of the Work and the supply of specified material and equipment. So long as Contractor awards a Subcontract to a pre-approved subcontractor or vendor for a portion of the Work or for the supply of materials or equipment for which such subcontractor or vendor has been pre-approved, no further approval by Owner is required before entering into such Subcontract. Contractor shall obtain Owner's approval in advance for all other Subcontracts by providing Owner information regarding the subcontractor or vendor with whom Contractor desires to enter into a Subcontract, the portion of the Work or the material or equipment to be furnished by such subcontractor or vendor and the reasons for such selection. Owner shall have ten (10) days after receipt of the aforementioned information from Contractor to approve or disapprove such subcontractor or vendor. If Owner fails to timely notify Contractor of its disapproval within such ten (10) day period, Owner shall be deemed to have waived its right to disapprove such subcontractor or vendor and if Owner subsequently requests a change or removal of any such subcontractor or vendor for reasons other than performance by such subcontractor or vendor which is not in accordance with the terms and conditions of this Agreement, Owner shall issue a Change Order appropriately adjusting the Maximum Contract Price and any affected Milestone Dates. Within five (5) days of the (i) the Effective Date, for those Subcontracts entered into prior to the Effective Date; or (ii) the date on which each Subcontract is subsequently signed, Contractor shall provide Owner with a copy of the Subcontract. Owner shall be designated as a third party beneficiary under each Subcontract, which Subcontract shall comply with the provisions of Section 3.3. No approval or disapproval by Owner shall relieve Contractor of its obligations to properly perform the Work in accordance with the terms and conditions contained in this Agreement. Neither the use or negligence of any subcontractor, vendor, materialman, or other supplier, nor the breach of any obligation of any subcontractor, vendor, materialman or supplier shall relieve Contractor of its obligation to properly perform the Work in a timely manner in accordance with the terms and conditions of this Agreement.

3.2.   Units. Owner has purchased four (4) Units and Associated Equipment which were dismantled by Contractor. When refurbishment is deemed satisfactory, the Units and Associated Equipment shall be shipped to the Site or to offsite storage by Contractor for receiving and offloading by Contractor. Contractor shall coordinate design and performance of the Work with the activities related to the Units, and shall, on behalf of the Owner, coordinate the delivery of the Units for transportation and delivery to offsite storage and to the Site. Contractor shall contract for and shall be solely responsible for the dismantling, loading and arranging shipment (with the costs of shipment to be reimbursed by Owner and not credited against the Maximum Contract Price) of the Units from _____ to _____ for refurbishing by Owner at Owner's expense and for receiving and offloading the Units and Associated Equipment at offsite storage and the Site.

3.3.   Subcontract and Supply Contract Terms. Contractor shall not enter into any Subcontract or supply contract with respect to the Work unless such Subcontract or supply contract contains at least the terms and conditions set forth below. Furthermore, in addition to providing a copy of each Subcontract to owner as provided in Section 3.1, Contractor shall maintain fully executed copies of all Subcontracts and supply contracts entered into with respect to the Work in its records and shall permit Owner to obtain complete copies of and to review such Subcontracts and supply contracts to determine if such Subcontracts and supply contracts contain the terms set forth below:

- The right of Owner to assign the Subcontract or supply contract, without having to first obtain the consent of the subcontractor or supplier, as security for any financing incurred in connection with financing the cost of constructing the Plant.

- The right of Owner or Owner's lenders, if any, to enforce rights under the Subcontract or supply contract if Contractor defaults under this Agreement.

- The right of Owner to enforce all warranties, whether express or implied, given to the Contractor by the subcontractor or supplier.

- All warranties of such subcontractor's work or supplier's materials or equipment shall be for a period equal to (i) eighteen (18) months from the date construction work is completed, equipment is first installed and placed in operation or materials are purchased and delivered to the Site; or (ii) twelve (12) months from the Plant Completion Date, whichever is the first to expire.

- The duty of the subcontractor or supplier to maintain confidentiality of Owner's proprietary information, and to execute and deliver lien waivers upon receipt of payment.

- As it relates to the performance and quality of work and materials to be provided by such subcontractor or supplier, the obligation of the subcontractor or supplier to abide by all of the conditions and obligations imposed upon Contractor under this Agreement.

- The obligation of the subcontractor or supplier to permit Owner or its representatives to have free access to those portions of the Work being performed or provided by the subcontractor or supplier (whether at the Site or elsewhere).

- The obligation of the subcontractor or supplier to comply with all of the polices and procedures of Owner with respect to the Site, including without limitation construction gate access, parking and storage of vehicles, equipment and materials, and labor, union, health and safety rules and regulations.

- The right of Contractor to terminate or suspend the Subcontract or supply contract for convenience.

ARTICLE IV

COMMENCEMENT AND COMPLETION OF WORK

4.1.   Commencement and Completion of Construction. Contractor has prior to the Effective Date proceeded with the procurement of certain equipment and materials and certain engineering and design services required to construct the Plant described in the TSD, and Contractor shall complete the engineering and design services with due diligence. All work performed by Contractor in connection with the Plant prior to the Effective Date and used by Contractor in the performance of the Work shall be deemed to have been performed pursuant to this Agreement, shall be subject to the terms and conditions hereof and payments made by Owner for such work shall be credited to the extent specified in Section 6.1 against the Maximum Contract Price. Contractor shall commence construction of the Plant at the Site by mobilization within _____ days after Owner gives Contractor written notice authoring Contractor to commence on-Site construction (the "Notice to Proceed with Construction Commencement"), and thereafter diligently pursue completion of the Work. Contractor agrees to complete the Work so as to meet the Milestone Dates, as such dates may be extended pursuant to the terms of this Agreement.

4.2.   Unit Mechanical Completion.

(a)   Mechanical completion, start-up and commissioning for each Unit shall occur separately for each Unit as work on each Unit is completed. Contractor agrees to achieve Unit Mechanical Completion of all four (4) Units on or before the Guaranteed Mechanical Completion Date, which date may be extended pursuant to the terms of this Agreement. A Unit shall have attained "Unit Mechanical Completion" when such Unit has been constructed and assembled at the Site in accordance with the design and engineering plans, specifications and drawings contained in the TSD, and all applicable codes, laws, rules and regulations, the fit of interconnecting components and proper rotation of motors has been confirmed, all necessary cleanout and flushes have been completed, all hydrostatic and pneumatic pressure tests, continuity of circuits and ground fault tests and other mechanical and electrical tests which are customarily conducted prior to the Start-up of equipment similar to the Unit have be completed, the Unit has been determined ready for operation to the extent that deficiencies which can be determined prior to the introduction of fuel and other materials have been corrected by the Contractor and the Unit has been determined to be capable of Unit Initial Commercial Operation. At such time as Contractor believes any Unit has achieved Unit Mechanical Completion, Contractor shall advise Owner in writing (each a "Unit Mechanical Completion Notice"), and Owner shall thereafter have _____ days following receipt of the Unit Mechanical Completion Notice to inspect the Unit, its Associated Equipment and BOP and notify Contractor in writing of any deficiencies in construction which are required to be corrected and/or completed in order for the Unit to have attained Unit Mechanical Completion ("Deficiency Notice"). Contractor shall use its "best efforts" to correct the deficiencies described in the Deficiency Notice within _____ days of receipt thereof. The Unit shall not be deemed to have attained Unit Mechanical Completion until all deficiencies listed in the Deficiency Notice, if any, are corrected by Contractor to Owner's reasonable satisfaction, or if no deficiencies exist, until Owner completes its inspection of the Unit following receipt of the Unit Mechanical Completion Notice and the Contractor has delivered to Owner the Turnover Documents. Owner shall notify Contractor in writing of its concurrence that the Unit has attained Unit Mechanical Completion, and sign a written notice of acceptance thereof, and deliver the same to the Contractor ("Acceptance of Unit Mechanical Completion"). Owner's concurrence that the Unit has attained Unit Mechanical Completion shall not be unreasonably withheld.

(b)   At such time as Owner concurs in writing that a Unit has attained Unit Mechanical Completion, Owner shall accept and assume, in writing, care, custody, and control of that Unit and its Associated Equipment and BOP, and title to such Unit and its Associated Equipment and BOP, subject to Owner's rights and Contractor's obligations under Section 4.3, 4.4, and 4.5 and Articles XI and XIV, and risk of loss thereto shall pass to Owner.

4.3.   Unit Start-up. Promptly following issuance by Owner of Acceptance of Unit Mechanical Completion for a Unit, Owner or Owner's operating subcontractor shall, under the supervision and direction of

Contractor, commence Start-up of that Unit ("Unit Start-up"). Contractor shall have one or more of Contractor's experienced technical personnel as well as technical representatives of each of the BOP equipment manufacturers present during each Unit Start-up to direct implementation of proper Start-up procedures and to assist with the resolution of any mechanical or other problems encountered with the Unit, the Associated Equipment or the BOP. Any and all expenses associated with having such technical personnel of Contractor and its BOP equipment manufacturers present for each Unit Start-up shall be credited against the Maximum Contract Price. In the event that a Unit is ready to commence Unit Start-up except for reasons solely attributable to Owner supplied equipment or services, then the affected Milestone Dates shall be adjusted on a day-for-day basis until Owner has corrected such deficiencies so that Unit Start-up can commence or resume. Contractor acknowledges that it has had access to the manufacturer's start-up and operations manuals for the Units and the Associated Equipment and therefore Contractor agrees that any and all problems related to monitoring and control of the Units and Associated Equipment shall be deemed to be the responsibility of Contractor until and unless Contractor is able to prove conclusively that the problem results from the mechanical condition of the Unit or the Associated Equipment instead of the control system supplied by Contractor.

(b)   Promptly following completion of Unit Start-up, Owner and Contractor shall jointly endeavor to run the Unit to demonstrate that it is capable of being started locally and run locally by two (2) or fewer on-Site personnel for eight (8) continuous hours at Fully Loaded Conditions with all Automatic Shutdowns in place and fully functioning to prevent potential damage to the Unit, its Associated Equipment or the BOP ("Unit Initial Commercial Operation"). Following successful demonstration of Unit Initial Commercial Operation of a Unit, Owner and Contractor shall jointly prepare a written Punch List of items related to that Unit which must be completed and/or corrected by Contractor, it being specifically understood that Contractor shall complete or correct all Punch List items within thirty (30) days after the Unit Initial Commercial Operation of that Unit.

4.4   Plant Commercial Operation and Performance Testing.

(a)   During the thirty (30) day period after the last of the Units has demonstrated that it is capable of Unit Initial Commercial Operation (the "Performance Testing Period"), Owner and Contractor shall jointly endeavor to run the Plant to prove that it has passed all tests required by applicable laws, ordinances, rules, regulations, orders and codes and the TSD (including any tests required by the utility into whose system the power generated by the Plant will be transmitted) and is capable of being remotely started, operated at nominal rated capacity to produce a net output equal to or greater than _____ MW and dispatched in the normal course of business ("Plant Commercial Operation"). During the Performance Testing Period Owner or Owner's operating subcontractor shall also test and verify the capability of (i) the materials and equipment sized and engineered by Contractor and installed as part of the Work and (ii) the Plant to otherwise operate in accordance with the performance criteria described in the TSD. Contractor shall correct any and all deficiencies in the Work performed by Contractor discovered by Owner during the Performance Testing Period at Contractor's sole cost and expense. Any deficiency in the Work discovered following completion of the Performance Testing Period shall be corrected by Contractor only to the extent that such deficiencies are the result of a breach of Contractor's guarantees and warranties under Article XI of this Agreement, and then under and pursuant to the terms and conditions contained in Article XI.

4.5   Final Acceptance. Upon (i) the Plant having attained Plant Commercial Operation, (ii) completion by Contractor of all items described in any Deficiency Notice or Punch List; (iii) satisfactory completion of the correction of all deficiencies in the Work discovered by Contractor during the Performance Testing Period; (iv) confirmation that all drawings (including as-built drawings), documents and other items which Contractor is required to provide pursuant to this Agreement have been provided to Owner; and (v) Owner's receipt of evidence in form satisfactory to Owner that there are no liens filed against the Work and all lien periods under _____ law have expired, Owner shall issue written notice of final acceptance of the Work ("Final Acceptance") to Contractor and the date of issuance of such written notice shall be the "Plant Completion Date."

Contractor's Incentive Fee, if any, and the remaining Retainage shall be paid by Owner to Contractor within thirty (30) days after Final Acceptance.

ARTICLE V

CONTRACT PRICE

5.1.   <u>Contract Price</u>. (a) Subject to the terms of this Article, as full and complete compensation to Contractor for (i) the satisfactory performance and lien-free completion of all of the Work, (ii) compliance with all of the terms and conditions contained in this Agreement by Contractor, and (iii) Contractor's payment of all obligations incurred in, or applicable to, the performance of the Work by Contractor, Owner shall pay to Contractor the sum of (w) Contractor's Costs plus (x) Contractor's Fees, (y) Contractor's Incentive Fees, if any, minus, if applicable (z) any Liquidated Damages and any amount reasonably determined by Owner to be necessary to correct any portion of the Work not performed in accordance with this Agreement; provided, however, that the total amount payable to Contractor before any deduction for the damages mentioned in (z) above, shall in no event exceed (the "Maximum Contract Price"). It is expressly understood and agreed that the Site is presently classified "Industrial Machinery," and Owner and Contractor have been advised that all of the Work is exempt from any sales and/or use taxes under current State of _____ tax laws and regulations. Owner agrees to timely apply for such exemption in its own name and to assist all of its major subcontractors in obtaining such exemption, and should any agency or authority assess or attempt to assess any sales and/or use tax against all or any part of the Work, Contractor shall promptly notify Owner and shall follow Owner's directions as to paying or contesting any such assessment. In the event that Owner directs Contractor to pay or contest such assessments, the actual expenses related to either of these directions will be reimbursed to Contractor by Owner.

(b)   The Maximum Contract Price is fixed and non-escalatable for the duration of the Work, and shall only be subject to adjustment to reflect changes in the Maximum Contract Price made pursuant to Article VIII. Contractor waives and releases any and all rights to claim additional compensation over and above the Maximum Contract Price based on "quantum merit" or other quasi contractual basis or in equity.

ARTICLE VI

SCHEDULE OF PAYMENTS

6.1.   Payment of Contract Price.

(a)   Payments made to the Contractor under Service Request No. 1 pertaining to the dismantling the Units and their Associated Equipment and Service Request No. 2 (except for that portion pertaining to the transportation and refurbishment of the Units and the Associated Equipment) and any other Service Requests which Owner may pay pertaining to Work performed prior to the Effective date shall be credited against the Maximum Contract Price. Contractor shall issue an invoice to Owner by the tenth (10th) day of each month for (i) Contractor's Costs and (ii) Contractor's Fees, incurred and earned during the prior month, such invoice to include documents and information so that Owner can confirm the correctness of each such invoice, partial lien waivers from each subcontractor, vendor, supplier and materialman the value of whose work included in such invoice exceeds $5,000, and a statement by Contractor that those items of the Work which are critical to achieving the Unit Mechanical Completion Date are on schedule.

(b)   Upon Owner's receipt of Contractor's invoice with supporting documentation, Owner shall conduct such review of the Work including, if deemed necessary by Owner, Contractor's records, to confirm the accuracy of such invoice. Unless Owner shall dispute in good faith any amount included in such invoice, Owner shall pay ninety percent (90%) of the amount of such invoice within thirty (30) days after receipt of Contractor's invoice, with the other ten percent (10%) (the "Retainage") being withheld and paid as provided in Sections 6.2 and 6.3 below. Payment of an invoice shall not (i) prejudice Owner's rights to object to any other invoice submitted by Contractor, (ii) constitute acceptance of any part of the Work by Owner or approval of any Contractor's Costs invoiced by Contractor, or (iii) nullify or modify in any manner the guarantees and warranties of the Contractor under Article XI.

(c)   If Owner disputes in good faith any amount included in such invoice, Owner shall notify Contractor of the amount which is in dispute and the reason for such dispute and shall pay ninety percent (90%) of the portion of such invoice not in dispute within thirty (30) days after receipt of Contractor's invoice. Owner has the absolute right to dispute in good faith and reject any invoice submitted by Contractor which is not in compliance with the requirements of this Agreement and may deny payment of any portion of the invoice (i) which is not supported by proper documentation and itemization of Contractor's Costs, (ii) which was incurred with respect to reworking portions of the Work to correct deficiencies in the Work, (iii) which relates to portions of the Work performed by subcontractors, vendors, suppliers or materialmen which have not provided Owner with partial lien waivers as required by Article VII, or (iv) if items of the Work which are critical to achieving the Mechanical Completion Date are behind schedule (in which case the amount withheld shall be equal to the Liquidated Damages applicable to the period by which such items of the Work are behind schedule). Contractor shall promptly take such actions as are necessary to correct the deficiency or deficiencies cited by Owner and upon such correction by Contractor, Owner shall pay ninety percent (90%) of the amount withheld by Owner related to each such deficiency within five (5) days after Contractor has established to Owner's reasonable satisfaction that the reason for each such dispute has been eliminated.

6.2   Unit Mechanical Completion Retainage Payment. When Owner issues a Unit Mechanical Completion Notice for any Unit, Contractor shall be entitled to invoice Owner and be paid fifty percent (50%) of twenty-five percent (25%) of the Retainage from the Retainage then accumulated.

6.3.   Final Payment. Following Final Acceptance, Contractor shall submit a final invoice for (i) Contractor's Costs incurred since the last monthly invoice, plus (ii) Contractor's Fees related to such Contractor's Costs, plus (iii) the amount the Retainage then being held by Owner minus, if applicable (iv) any Liquidated Damages and any amount reasonably determined by Owner to be necessary to correct any portion of the Work not performed in accordance with this Agreement, plus, if applicable (v) Contractor's Incentive Fee. Contractor shall also submit a Waiver of Liens Claim in the form attached hereto as Annex 3. Owner shall pay such invoice within thirty (30) days of receipt thereof.

Upon Owner's request, Contractor will furnish a written breakdown of all invoiced amounts by individual component of the facility so as to enable Owner to create tax depreciation schedules.

ARTICLE VII

INVOICING INSTRUCTIONS

One (1) original and two (2) copies of invoices, together with all other required documentation specified in Article VI shall be transmitted to:

_____

_____

_____

Attn: _____

To assist in the administration of this Agreement, separate invoices shall be issued for the base scope of Work and for each Change Order, each invoice shall specify the Maximum Contract Price, as it may have been adjusted to the date of invoice, the total amount invoiced to date, the total amount previously paid by Owner, the total amount of Retainage then held by Owner and the amount then due. The invoices for the base scope of Work and for Change Orders will be netted against each other during any given month and Owner shall remit the net amount to Contractor. Contractor shall provide a signed Release and Waiver of Claims in the form attached as Annex 3 as often as requested by Owner and with its final invoice. Contractor shall cause its subcontractors, vendors, suppliers and materialmen to execute and deliver Release and Waiver of Claims in the form attached as Annex 4 to Owner in exchange for Contractor's payment to such parties.

ARTICLE VIII

CHANGES IN THE WORK

8.1   Owner's Right to Order Changes.

(a)   Owner, without invalidating this Agreement, may order changes in, additions to and deletions from the Work from time to time, with all such changes to be ordered in writing. If such changes involve extra cost to Contractor or will adversely affect the Work or Contractor's ability to meet any Milestone Date, Contractor shall so advise Owner in writing prior to commencing performance of such change. Contractor's notice to Owner shall be in writing, be provided promptly following Owner's request for such change, and include such detail as Owner may reasonably request as to the effect of the requested change on the Maximum Contract Price, the Milestone Dates and the Work. If such notice is not so given, it shall be deemed that no change of the Maximum Contract Price, the Milestone Dates or the Work is required by reason of such change.

(b)   If such notice is given by Contractor, Owner and Contractor shall endeavor in good faith to agree upon an adjustment to the Maximum Contract Price, the Milestone Dates, and any other terms of this Agreement affected by such change. In the absence of such agreement, Owner shall have the right to direct Contractor to proceed with the change based on Owner's good faith assessment of the equitable price adjustment subject to later resolution of the adjustments to be made to the Maximum Contract Price, the Milestone Dates, and/or other affected provisions of this Agreement by mutual agreement of the parties or by arbitration under Article XXV. The adjustment to the Maximum Contract Price shall be determined in one or more of the following ways:

(i)   Mutual acceptance of a lump sum which is itemized and supported by substantiating data.

(ii)   Mutually agreed unit prices (with the actual increase or decrease in the Maximum Contract Price to be determined later).

(iii)   Time and material charges based on Contractor's current rate schedule and/or the rate schedule which Contractor has with its affected subcontractor(s) (with the actual increase or decrease in the Maximum Contract Price to be determined later).

(iv)   If the Change Order would result in a deduction from the Maximum Contract Price and the parties cannot agree to an amount as set forth in (i) to (iii) above, then Contractor and Owner shall each submit an estimate of the reduction for such Change Order and Owner shall select an independent third party who shall also submit an estimate of such reduction. Whichever of Contractor's or Owner's estimate is closest to the third party's estimate shall be the amount deducted. If Contractor's and Owner's estimates are equally close to the third party's estimate, then the third party's estimate shall be the amount of the deduction. If there is a fee associated with obtaining the third party estimate, the Contractor and Owner shall each pay fifty percent (50%) of such fee.

(c)   All changes to the Maximum Contract Price, the Milestone Dates, and/or the scope of the Work shall only be effective if made pursuant to a written change order (each, a "Change Order") executed as follows:

(i)   if the Change Order does not require an individual adjustment to the Maximum Contract Price greater than $_____, an adjustment to the Maximum Contract Price which, when aggregated with all previously approved adjustments to the Maximum Contract Price, establishes a Maximum Contract Price which is $_____ or more greater than the original Maximum Contract Price, an individual adjustment to any Milestone Date of more than _____ days or an adjustment to any Milestone Date which, when aggregated with all previously approved Milestone Date adjustments, establishes a Milestone Date which is _____ or more days after the original Milestone Date, then such Change Order shall be executed by Owner's principal representative and Contractor's Project Manager or

(ii)   if the Change Order requires an adjustment to the Maximum Contract Price of greater than $_____, an adjustment to the Maximum Contract Price which, when aggregated with all previously approved adjustments to the Maximum Contract Price, establishes a Maximum Contract Price which is $_____ or more greater than the original Maximum Contract Price, an individual adjustment to any Milestone Date of more than _____ days or an adjustment of any Milestone Date which, when aggregated with all previously approved Milestone Date adjustments, establishes a Milestone Date which is or more days after the original Milestone Date, then such Change Order shall be executed by four individuals, Owner's principal representative, a vice president or president of Owner, Contractor's Project Manager, and a vice president or president of Contractor.

A Change Order shall also be promptly executed, as provided above, to reflect any decision rendered in any arbitration proceeding conducted pursuant to Article XXV.

8.2   Contractor's Claims.

(a)   If Contractor claims that any instruction, act or omission of Owner or any of Owner's representatives constitutes a change in the Work because it causes the Contractor to incur additional cost and/or delays the performance of all or any part of the Work, then Contractor shall promptly give written notice to Owner specifying the instruction, act or omission which gave rise to the claim and an explanation as to how such instruction, act or omission has affected the Work, including the effect on Contractor's Costs , the Milestone Dates and any other affected provision of the Agreement. Provided that Owner determines that Contractor's claim is reasonable, is supported by clear evidence and Contractor has given prompt notice as provided above, then Owner shall issue a Change Order adjusting the Maximum Contract Price, the Milestone Dates and any other affected provision of the Agreement.

(b)   If Contractor incurs costs of reconstruction and/or costs to replace and/or repair damaged equipment, materials or supplies for which Contractor is not compensated by the proceeds of applicable insurance or otherwise, and such damage resulted from the fault or neglect of Owner or any of Owner's representatives or Owner's contractor separately employed by Owner or any so-called "war risk," then Contractor shall promptly give written notice to Owner specifying the cause of such damage and an explanation of the effect of such damage on Contractor's Costs, the Milestone Dates and any other affected provision of this Agreement. Provided that Owner determines that Contractor's claim is reasonable, is supported by clear evidence and Contractor has given prompt notice as provided above, then Owner shall issue a Change Order adjusting the Maximum Contract Price, the Milestone Dates and any other affected provision of the Agreement.

(c)   Notwithstanding anything to the contrary set forth herein, no Change Order shall be issued and the Maximum Contract Price, the Milestone Dates and any other affected provisions of the Agreement shall not be adjusted if and to the extent that such claim is attributable to any act or omission of Contractor, or any if its representatives, subcontractors, vendors, materialmen or suppliers.

8.3   <u>Changes to the Terms and Conditions of this Agreement</u>. All changes to any provisions of this Agreement (other than those provisions related to the scope of the Work and changes to the Maximum Contract Price, the Milestone Dates and other affected provisions directly resulting from changes to the scope of Work) shall only be effective if made in an amendment to this Agreement signed on behalf of Owner by a vice president or president of Owner and on behalf of Contractor by a vice president or president of Contractor, or as directed by arbitration under Article XXV.

8.4   <u>Contractor's Right to Reimbursement for Evaluating Changes</u>. In the event that Owner has ordered a change or has requested that Contractor evaluate the effects of a proposed change, Contractor shall include its costs for evaluating the effects of the change or the proposed change in the Maximum Contract Price adjustment quoted by Contractor in its notice to Owner concerning such change. If Owner and Contractor agree on such change and sign a Change Order or an amendment to this Agreement reflecting such change, or if the change in the Maximum Contract Price is determined by arbitration under Article XXIV, the adjustment made to the Maximum Contract Price, if any, stated in such amendment or arbitration award shall be deemed to include Contractor's Costs for evaluating the effects of such change. If Owner decides not to proceed with such change, Owner shall reimburse Contractor its actual costs of evaluating the effects of such change at the time of payment of Contractor's next invoice.

ARTICLE IX

INSPECTION AND REJECTION OF MATERIALS AND WORKMANSHIP

9.1.   General. During the performance of the Work performed at the Site, Contractor shall follow and comply with any special procedures provided by Owner in order to protect and preserve Owner's existing property located on the Site and the personnel on the Site in connection with such existing property. The Work is subject to inspection and testing by Owner at all reasonable times and at any and all places where any part of the Work is being performed, including at the facilities of subcontractors, vendors, materialmen and suppliers. Owner shall designate items of Work which it specifically desires to inspect and test and Contractor shall provide Owner advance notice of readiness for inspection and testing of all such items. Failure of the Owner to inspect or test or to discover defects or to object thereto shall not prejudice or operate as a release or waiver of any rights, including the right to inspect or reject at a later time, nor shall it release Contractor from or modify any guaranty, warranty or other responsibility under this Agreement. Contractor shall furnish, at its expense, such facilities as may be necessary for conducting any such inspection and tests clearly identified and agreed upon as requirements under this Agreement (including Exhibits and Annexes). Facilities that are required for Owner to conduct any additional inspections or tests will be provided by Owner. If Contractor has not given proper prior notice of readiness for inspection as required above and such Work is covered or otherwise made inaccessible before such inspection is made without Owner's consent, Contractor shall bear the expense of uncovering and recovering such Work so that Owner may inspect such Work.

9.2.   Owner's Right to Order Uncovering of Work. Owner shall have the right to order the uncovering of Work whether or not subject to prior inspection hereunder, whether or not actually inspected by Owner, and whether or not Owner had previously waived such inspection. Except as otherwise provided in Section 9.1, Owner shall bear the reasonable direct cost of uncovering and redoing the Work, and, should such action also delay critical path project activities, the applicable Milestone Dates shall be extended based the number of days lost as a result of the uncovering and recovering of the affected Work unless any defect in excess of those allowed by the applicable codes and standards or non-compliance with this Agreement is found, in which case all costs of uncovering, correcting and recovering the Work shall be borne by Contractor, and the Milestone Dates shall not be extended.

ARTICLE X

LIQUIDATED DAMAGES

10.1.   Delay Damages. (a) If any Unit has not attained Unit Mechanical Completion by the Guaranteed Unit Mechanical Completion Date or if the Plant has not attained Final Acceptance by the Guaranteed Plant Completion Date, as such dates may have been extended pursuant to the terms of this Agreement, then Owner will suffer damages which will be difficult to quantify precisely, including the loss of future profits and possible loss of contractual rights accruing to Owner under certain other agreements entered into on the date hereof with certain other parties in connection with the Plant. Accordingly, Owner and Contractor agree that Owner may, at the time of final payment, offset and deduct from the final payment, as fixed, agreed and liquidated damages, and not as a penalty, the amount of _____ per Unit per day for each day after the Guaranteed Unit Mechanical Completion Date that Unit Mechanical Completion has not been attained and the amount of _____ Dollars ($ _____) per day for each day after the Guaranteed Plant Completion Date that Final Acceptance has not been attained ("Liquidated Damages"). In no event shall Contractor be liable to Owner for Liquidated Damages in a cumulative amount greater than _____. It is specifically understood and agreed by Owner and Contractor that the payment of Liquidated Damages, calculated and paid as provided above, shall be in lieu of and shall be deemed to be full compensation by Contractor to Owner for and in respect of any delay damages actually incurred by Owner.

10.2   Owner's Other Rights. Notwithstanding the agreement between Contractor and Owner with respect to the payment of Liquidated Damages as provided in Section 10.1, at any time after Contractor's cumulative liability to Owner is equal to or greater than _____ Owner shall be entitled to discontinue the accrual of further Liquidated Damages and instead exercise Owner's other rights under Section 20.1. If Owner exercises its rights under Section 20.1, Owner shall continue to be entitled to offset and deduct any Liquidated Damages previously accrued in accordance with Section 10.1. The maximum limitation on Liquidated Damages established under Section 10.1 does not constitute a limitation on Owner's rights against Contractor under any other provision of this Agreement.

ARTICLE XI

CONTRACTOR GUARANTEES AND WARRANTIES

11.1.   Scope of the Warranty. It is understood and agreed that Owner is providing the Units and Associated Equipment and Contractor is making no performance guarantees or warranties with respect to the Units or Associated Equipment. Contractor's guarantees and warranties with respect to the Work shall be as follows:

(a)   Engineering Design. Contractor guarantees and warrants that the detailed engineering design to be performed by or on behalf of Contractor hereunder shall be in accordance with accepted engineering practices and shall conform to all applicable codes, laws, rules and regulations and the design basis furnished to Owner by Contractor in the TSD. Contractor shall, all at its sole cost and expense, make such changes or modifications in the design of the Work, and repair, modify, and/or replace all or any portion of the Work which will not operate in accordance with the engineering design parameters or which fails as a result of Contractor's defective design of, or defective field workmanship on, the Work. If correction of such defective design or field workmanship requires changes or modifications to the materials and equipment installed and in place at the Plant, Contractor shall bear all costs and expenses incurred in connection with such changes or modifications, including without limitation, the repair and replacement of materials and equipment.

(b)   Material and Equipment. Contractor shall, for the protection of Owner, secure from all of its vendors, suppliers and materialmen guarantees and warranties with respect to the material and equipment purchased by Contractor and manufactured and/or supplied by such vendors, suppliers and materialmen as a part of the Work that such materials and equipment are new and unused (unless the TSD specifically provides for the use of used materials and equipment or the use of used materials and equipment have been specifically approved in writing by the Owner), shall not fail under design operating conditions, and shall be free from defects in design, engineering, workmanship and material for a period of (i) eighteen (18) months from the date the equipment is first installed and placed in operation or materials are purchased and delivered to the Site; or (ii) twelve (12) months from the Plant Completion Date, whichever is the first to expire, but the foregoing warranty does not guarantee against the failure of or damage to the Work resulting from a lack of normal maintenance or as a result of changes or additions to the Work made or performed by persons not directly responsible to Contractor except where such changes or additions are made by such persons in accordance with Contractor's directions. Contractor shall take all actions necessary to ensure that the guarantees and warranties shall be directly enforceable by Owner and shall obligate the vendor, supplier or materialman to repair and/or replace any defective materials or equipment. In addition to obtaining the required guarantees and warranties from vendors of material and equipment, Contractor itself guarantees and warrants that the material and equipment fabricated by Contractor, are new and unused (except as provided above), shall not fail under design operating conditions, and shall be free from defects in design, engineering, workmanship and material for a period of (i) eighteen (18) months from the date the equipment is first installed and placed in operation or materials are purchased and delivered to the Site; or (ii) twelve (12) months from the Plant Completion Date, whichever is the first to expire.

(c)   Workmanship. Contractor warrants that all Work and services performed by Contractor or its subcontractors under this Agreement will be performed in a first-class, skillful and workmanlike manner, in accordance with the requirements of this Agreement and all Annexes and Exhibits attached hereto, and in a manner consistent with accepted engineering and construction practices prevailing for such type of work in the electric power industry. Contractor's warranty of workmanship shall also apply to all items of equipment manufactured or fabricated by Contractor or by subcontractors of Contractor.

11.2.   <u>Remedies for Breach of Warranty</u>. Owner's remedies for breach of the guarantees and warranties contained in Section 11.1 shall be as follows:

(a)   For breach of the guarantees and warranties made in Section 11.1(a), Contractor shall re-perform, at Contractor's expense and without limit, that portion of the engineering design for which such breach has occurred, and provided that any claim or demand shall be made by Owner in writing prior to twelve (12) months after the Plant Completion Date. In addition, Contractor shall bear the total cost of material or equipment to be replaced and the installation cost required as a result of defects, errors or omissions in the engineering design, provided that Contractor's responsibility is limited to the scope of Work defined by this Agreement, and provided further that any claim or demand shall be made by Owner in writing prior to twelve (12) months after the Plant Completion Date.

(b)

(i)   For breach of the other guarantees and warranties provided by Contractor in Section 11.1(b), Contractor shall repair or replace, or cause to have repaired or replaced under its supervision, at its own expense and without limit, at the Site, any faulty or defective workmanship, materials or equipment covered by the guarantees and warranties set forth in Section 11.1(b) which is shown to be faulty or defective within (i) eighteen (18) months from the date such equipment was first installed and placed in operation or material was purchased and delivered to the Site; or (ii) twelve (12) months from the Plant Completion Date, whichever is the first to expire. Owner shall promptly notify Contractor of such fault or defect upon discovery in writing, and upon receipt of this written notice, Contractor agrees to assess the claim, submit recommended resolutions, and perform the mutually agreed upon resolutions. All of the foregoing steps shall be taken as soon as is reasonably practical, giving due consideration to Owner's operating schedule (including, without limitation, performance of corrective actions on Saturdays, Sundays, holidays, and after normal working hours). Such defects shall be exclusive of those incurred because of ordinary wear and tear, corrosion or erosion (not attributable to Contractor's failure to use accepted engineering practice) or improper maintenance or operating conditions more severe than those contemplated by the original design of the Plant. Contractor's responsibility for guarantee and warranty obligations shall include repair and/or replacement of any defective materials covered by the warranties and guarantees set forth in Section 11.1(b), freight charges both to and from the job site, labor for disassembly and reassembly of affected parts and all attachments in the immediate area, and routine functional testing to ensure that the original guarantee has been fully satisfied. If Contractor, after receiving the aforesaid notice, fails or refuses to repair or replace such defective materials, equipment, or workmanship promptly, Owner may repair or replace such defective material or equipment and/or correct such defective workmanship at Owner's expense and such expense shall be deducted from any then unpaid portions of Contractor invoices or shall be invoiced to Contractor accordingly. In all events, Contractor shall take all reasonable steps, short of litigation, to assist Owner in the enforcement of vendor, supplier and materialmen guarantees and warranties covering all items of materials and equipment. In addition, if the vendor's, supplier's or materialman's guarantee and warranty of an item of material or equipment has not expired by the time the warranty period under this Article XI expires, Contractor shall assign such guarantee or warranty to Owner at such time and, if necessary, continue to assist Owner in the enforcement of such guarantee and warranty until its expiration.

(ii)   Contractor's liability under Section 11 .2(b)(i) shall be limited as follows: Contractor shall have no guarantee or warranty obligation with respect to the Units other than that the Units are in the same condition as the Units were delivered to Contractor and that the Units are installed in accordance with the engineering design approved by the Owner and all recommendations of the manufacturer of the Units.

(c)   If Contractor takes any corrective action pursuant to Section 11.2(b), Contractor shall rewarrant all materials and equipment which are repaired or replaced for an additional period equal to (i) twelve (12) months after the date of the completion of such repair or replacement or (ii) the date on which Contractor's guarantees and warranties contained in Section 11.1(b) would otherwise have expired,, whichever is first to occur.

11.3   No Waiver. The duties, liabilities, and obligations of Contractor as set forth in this Article XI shall not be deemed waived, released, or relieved by Owner's inspection of, approval of, or payments made to Contractor for any portion of Contractor's work and services hereunder.

11.4   Exclusive Remedies. The above described remedies are the Owner's exclusive remedies for breach of this warranty unless such breach gives rise to an indemnity claim under Article XIV, or claims arising out of fraud and/or intentional misrepresentation.

11.5   No Other Warranties. Contractor makes no guarantee or warranty as to the quality of the Work, express or implied, except as set forth in this Article XI and in Section 12.1.

ARTICLE XII

COMPLIANCE WITH LAWS AND REGULATIONS; PERMITS

12.1.   Occupational Safety and Health Act. Contractor hereby warrants that all of the Work shall be performed and the Plant as installed shall be capable of being operated in compliance with the Occupational Safety and Health Act (1970), as amended to the Effective Date, and regulations promulgated thereunder (collectively, "OSHA"). If subsequent to such date there are any changes in OSHA, Contractor shall use its best efforts to identify the applicability of such changes to the Plant and to evaluate alternate means by which the Plant may be brought into compliance with such changes and the effect on the Maximum Contract Price, the Milestone Dates and other affected provisions of this Agreement of each alternative means of achieving compliance. Any change to the Work ordered by Owner as a result of changes in OSHA shall be treated as a change pursuant to Article VIII.

12.2.   Other Laws.

(a)   In the performance of the Work, Contractor at all times shall comply with and shall indemnify and hold Owner harmless from and against all costs, damages and expenses arising out of or resulting from any actual or alleged violation of any laws, ordinances, rules, regulations and orders, whether federal, state or local, (i) applicable to Contractor as an entity including, but not limited to laws, rules and regulations related to qualification to do business, equal employment opportunity, hours of employment, wage rates, unemployment compensation, workers' compensation and social security laws, and the maintenance of Contractor's facilities in compliance with OSHA and other laws related to safety in the workplace or protection of the environment, and (ii) applicable to the Work including, but not limited to laws, ordinances, rules, regulations and orders related to the safety of construction sites and building, piping and wiring codes. Contractor shall file all reports, pay all taxes, fees and charges required by such laws, rules and regulations. Contractor hereby certifies that it is and will continue to be in compliance with the "Fair Labor Standards Act of 1938," as amended, and all invoices shall so certify.

(b)   Without in any way limiting the responsibilities of Contractor set forth above, if the Work or the TSD are at variance in any respect with any laws, ordinances, rules, regulations or orders, Contractor shall promptly notify Owner in writing and shall promptly modify the Work and/or the TSD so that the Work and the TSD are in compliance with such laws, ordinances, rules, regulations and orders, it being specifically understood that Contractor shall not be entitled to submit a claim for a Change Order under Section 8.2 for an adjustment to the Maximum Contract Price and/or the Milestone Dates as a result of any such variance unless such variance results from a change in such laws, ordinances, rules, regulations or orders occurring after the Effective Date. If Contractor performs any Work contrary to such laws, ordinances, rules, regulations or orders, Contractor shall be fully responsible therefore and shall bear all costs attributable thereto.

12.3. <u>Plant Procedures</u>. Contractor when at the Site shall at all times comply and assure compliance by its subcontractors, vendors, materialmen and suppliers with Owner's procedures and policies (including, but not limited to, those related to safety and the protection of the environment), as those procedures and policies may be updated from time to time by written notice from Owner to Contractor.

12.4. <u>Obtaining Permits</u>. Owner, with the assistance of Contractor, shall be responsible for and shall obtain, at Owner's expense, the environmental permits required for the operation of the Plant and Contractor, with the assistance of Owner, shall obtain all other licenses and permits required or allowed to be in Contractor's name including, without limitation, the building permit and all other permits needed during construction which are required for Contractor to perform the Work, according to the terms of this Agreement and the TSD.

ARTICLE XIII

DISCLAIMER OF CONSEQUENTIAL DAMAGES

Except as otherwise provided for herein, regardless of whether recovery is sought under contract, tort (including negligence), strict liability or other legal theory, neither party shall be liable to the other party for any special, incidental, indirect, or consequential damages of any nature, including, but not limited to, loss of use of the Plant, downtime, loss of operating supplies, cost of money or loss of revenues, profits, or income. The foregoing limitation shall not, however, in any way limit Contractor's liability for Liquidated Damages under Article X or its indemnity liability under Article XIV for injuries or damages to persons or property.

ARTICLE XIV

INDEMNITY

Contractor hereby agrees and obligates itself to indemnify, defend and hold harmless Owner, its parent and affiliated companies, its lenders, and its and its lenders independent engineers, and the agents, officers, directors and employees of any of them (collectively, the "Indemnified Parties"), from and against any and all claims, demands, causes of action, suits, liabilities, judgments, damages, losses, interest, costs, expenses, and attorneys' fees, arising out of or resulting from any actual or alleged bodily injury or sickness (including death at anytime resulting therefrom) suffered by any person or persons whomsoever including, but not limited to, agents, officers, directors or employees of Contractor or any subcontractor, vendor, materialman or supplier of Contractor, and damage to or destruction of any property whatsoever including, but not limited to, the Work, to the extent that such claim, demand, cause of action, suit, liability, judgment, damage, loss, interest, cost, expense or attorneys' fees arise out of or result from, in whole or in part, or directly or indirectly, the negligence of Contractor, its subcontractors, vendors, materialmen, suppliers or the agents, affiliates, officers, directors or employees of any of them, excluding only claims, demands, causes of action, suits, liabilities, costs, expenses, or attorneys' fees based on the sole negligence or the gross negligence or willful misconduct of an Indemnified Party. The contributory negligence of any other party shall not be a defense to or limitation of the enforceability of Contractor's indemnity obligation under this Article.

Nothing herein is intended to or shall modify, affect or impair the protection afforded to Contractor under the All Risk Builder's Risk Insurance carried by Contractor for the benefit of both Owner and Contractor, it being specifically agreed and understood that Contractor shall not be responsible or liable for any damage to or destruction of the Work if Owner receives full compensation for such damages or destruction from the aforesaid insurance. Further, it is specifically understood that the liability of Contractor under this Article XIV shall not exceed the type and limits of or risks covered by the insurance required to be carried by Contractor under the provisions of Article XV.

ARTICLE XV

INSURANCE

15.1.   Contractor's Liability Insurance. Contractor, prior to the commencement of any Work under this Agreement, shall purchase and, until completion of the Work, maintain such insurance as will protect it and Owner and the other Indemnified Parties from and against those claims, demands, causes of action, suits, liabilities and judgments, damages, losses, interest, costs, expenses, and attorneys' fees which may arise out of or result from Contractor's operations under this Agreement or the performance of the Work by any subcontractor, vendor, materialman or supplier of Contractor. All such insurance shall be issued by companies licensed to do business in the appropriate states and having a general policy holder's rating of "A" and a financial rating of XII or better in the most recent edition of Best's Insurance Guide. If Contractor subcontracts or enters into any contractual arrangement with any subcontractor, vendor, materialman or supplier which does not carry similar insurance, Contractor shall assume the risk of such subcontractor, vendor, materialman or supplier not having such insurance.

The insurance to be purchased and maintained by Contractor hereunder shall be written for not less than the amount provided hereinafter or required by law, whichever is greater, said insurance to include, without limitation, the following:

(a)   insurance for liability under the workers' compensation or occupational disease laws of any state or other jurisdiction in which any part of the Work under this Agreement is performed (or be a qualified self insurer in those states and jurisdictions) or otherwise providing:

>   (i)   coverage for the statutory limits of all claims under the applicable State Workers' Compensation Act or Acts; the Jones Act or the Federal Employer's Liability Act; the Longshoreman and Harbor Workers' Compensation Act; (or comparable laws in other jurisdictions); and

>   (ii)   Employer's Liability Insurance with a single limit of not less than $_____.

(b)   Commercial General Liability Insurance with limits of not less than $_____ combined single limit each occurrence and in the aggregate for Products liability/completed operations, $_____ combined single limit per occurrence for Bodily Injury and Property Damage (excluding the Work covered by this Agreement) and $_____ general aggregate limit; such insurance to include blanket contractual coverage, broad form property damage, and, where applicable, the x, c, and u exclusions deleted with respect to the Work.

(c)   Comprehensive Automobile Liability Insurance covering all owned, hired and non-owned automotive equipment used in connection with the Work, with a combined single limit of not less than $_____ for each occurrence involving injury and/or property damage.

(d)   Excess Liability Insurance in umbrella form, which coverage shall be excess to the Commercial General Liability Insurance, Comprehensive Automobile Liability Insurance and Employer's Liability Insurance noted above, in an amount of not less than $ each occurrence and in the aggregate, where applicable.

(e)   Each of the liability insurance policies referenced under (b), (c) and (d) above shall be endorsed to name the Indemnified Parties as additional insureds with respect to any liability imputed to the Indemnified Parties arising out of or resulting in whole or in part from Contractor's or its subcontractors', vendors', materialmen's or suppliers' performance of the Work and waiving all subrogation rights against the Indemnified Parties.

(f)   Marine "Open Cargo" Insurance insuring all materials and equipment shipped by vessel against loss or damage arising from customary "all risk" marine perils while in transit, including a provision for payment of expediting expenses in the event of loss.

(g)   All Risk Builders Risk Insurance in an amount equal to the full value of the Work at risk from time to time, such insurance to provide coverage until Final Acceptance, and to name Owner, its lenders, Contractor and all subcontractors of any tier as insureds and as loss payees, as their interests may appear and waiving subrogation rights against the Indemnified Parties.

(h)   Certificates of Insurance acceptable to Owner shall be provided to Owner prior to the commencement of the Work. These Certificates shall contain a provision that coverages afforded under the policies shall not be canceled or modified until at least thirty (30) days prior to written notice, ten (10) days for non-payment of premium, has been given to Owner. At

Owner's request, Contractor shall also provide copies of its actual insurance policies to Owner for Owner's review.

ARTICLE XVI

[INTENTIONALLY OMITTED]

ARTICLE XVII

TITLE AND RISK OF LOSS

Subject to Contractor's warranty obligations under Article XI and its indemnity obligations under Article XIV, title to and all risks of loss of or damage to the Work and all materials, equipment, supplies and other things which are deliverable by Contractor pursuant to this Agreement for each Unit shall pass to Owner on Unit Mechanical Completion and title and all risks of loss of or damage to those materials, equipment, supplies and other things which are not directly related to a particular Unit shall pass to Owner on Unit Mechanical Completion of the last Unit; provided however, at Owner's election, Owner shall have the right at any time by written notice to Contractor to have title to all materials, equipment, supplies and other things for which Owner has paid Contractor pursuant to Article VI immediately pass to Owner and upon any such election, Contractor shall execute appropriate documents as reasonably requested by Owner to evidence Owner's ownership and Contractor agrees to affix labels or other tangible indication of Owner's ownership to all such materials, equipment, supplies and other things to the maximum extent practicable. Notwithstanding that Owner may elect to have title pass prior to Unit Mechanical Completion, unless Owner's written notice to Contractor shall specify otherwise, all risks of loss and damage to the Work and all materials, equipment, supplies and other things which are deliverable by Contractor pursuant to this Agreement shall remain with Contractor and pass at the time specified above.

ARTICLE XVIII

FORCE MAJEURE

18.1.   Definition of Force Majeure. For the purposes hereof, "force majeure" shall be any event, occurrence, happening or condition beyond the reasonable control of the party affected, including, but not limited to, the following: acts of God or the public enemy; acts of war, serious public disorder, rebellion, terrorism, or sabotage; floods, hurricanes, tornadoes, or other weather related events which cause actual delay to the Contractor in excess of the thirty (30) Rain Days included in the Project Schedule; strikes, labor disputes, whether direct or indirect; inability to obtain labor or materials, not due to delay or negligence in placing orders in a timely fashion and not due to the failure of any subcontractor or supplier selected by Contractor to fulfill its contractual obligations; and, for the Contractor, the failure of Owner to cause the last of the Units to be ready for shipment to the Site or offsite storage for receipt and offloading by Contractor by the date which is thirty (30) days after the issuance of the Notice to Proceed with Construction Commencement.

18.2.   Effect of Force Majeure. Delay in or failure of performance of either party shall not constitute a default hereunder, or be the basis for, or give rise to, any claim for damages, if and to the extent such delay or failure is caused by force majeure and the time for performance of an obligation (including the Milestone Dates), other than the obligation of Owner to make payments to Contractor pursuant to Article VI, shall be extended by a period of time equal to the time lost by reason of the delay which could not be overcome through the use of the best efforts of the party affected by the force majeure; provide, however, if Contractor experiences more than thirty (30) Rain Days during the performance of the Work, the Milestone Dates shall be extended on a day-for-day basis up to a maximum of seventeen (17) days.

18.3.   Notice of Force Majeure. The party who is prevented from performing by force majeure shall be obligated to give notice to the other party promptly after the occurrence or detection of such event, such notice to set forth in reasonable detail the nature thereof and the anticipated extent of the delay, and to use its best efforts to remedy such cause as soon as reasonably possible; provided, however, that the settlement of strikes, lockouts, and other industrial disturbances shall be entirely within the discretion of the party involved shall not be required to make settlement of strikes, lockouts, and other industrial disturbances by acceding to the demands of the opposing party or parties when such course is, in the judgment of the party involved, unfavorable to such party.

18.4.   <u>Owner's Right to Terminate for Extended Force Majeure</u>. In the event that the cumulative effect of one or more events of force majeure is to extend the Plant Completion Date by ninety (90) days or more, then at any time on or after such ninetieth day, Owner shall have the right to terminate this Agreement. If Owner elects to so terminate, then the provisions of Section 20.2 shall govern except that no profit shall be allowed Contractor if the reason for the termination is pursuant to this Section 18.4 and the event of force majeure is one affecting Contractor. If Owner terminates as the result of an event of force majeure affecting Owner, all provisions of Section 20.2 shall apply.

ARTICLE XIX

SUSPENSION OF WORK

Owner shall have the right at any time to order a suspension of all or any part of the Work by written notice to Contractor. Any such notice by Owner shall specify the portion of the Work to be suspended and the anticipated duration of such suspension, and Contractor shall promptly comply with such notice.

In the event such suspension is ordered by Owner because Contractor, after receiving appropriate notice from Owner, has continued its performance of the Work in a manner not in accordance with the requirements of this Agreement, such suspension shall not be deemed a change and no provisions of this Agreement including, but not limited to the Maximum Contract Price or any of the Milestone Dates, shall be adjusted as a result of such suspension. Promptly after Owner is satisfied that Contractor and Owner have reached agreement for resuming work in keeping with the requirements of the Agreement, or a reasonable plan for the correction of such non-conformance has been agreed upon, Owner shall issue a written notice to Contractor to resume the Work.

In the event such suspension is ordered for Owner's convenience, such suspension shall be treated as a change pursuant to Article VIII.

ARTICLE XX

TERMINATION OF AGREEMENT

20.1.   <u>Termination for Contractor's Breach</u>.

(a)   If insolvency, receivership, or bankruptcy proceedings shall be commenced by or against Contractor and not dismissed within thirty (30) days, or if Contractor shall make an assignment for the benefit of its creditors, or if Contractor shall fail or refuse to perform a material portion of the Work and Contractor does not commence to cure such failure or refusal within fifteen (15) days following written notice by Owner and thereafter diligently pursue such cure to completion, then this Agreement shall be deemed to have been breached by Contractor, and Owner, without prejudice to and cumulative of all other rights and remedies which Owner may have as the result of such breach, shall have the option to terminate this Agreement forthwith by giving written notice thereof to Contractor.

(b)   In the event of any such termination for Contractor's breach, Owner shall be entitled to enter upon Contractor's premises and take possession of and remove all drawings, specifications, materials, equipment and supplies intended to be incorporated into the Work; to have assigned to it all subcontracts and other contractual arrangements entered into by Contractor with any subcontractors, vendors, materialmen and suppliers for any part of the Work; and to make such arrangements as Owner in its sole discretion shall deem necessary or appropriate for the completion of the Work.

(c)   Contractor shall not be entitled to receive any further payment until the Work is finished. If the cost to Owner of finishing the Work exceeds the unpaid balance of the Maximum Contract Price, Contractor shall promptly pay the difference to Owner. If the cost to Owner of finishing the Work plus Owner's additional costs incurred for managerial and administrative services is less than the unpaid balance of the Maximum Contract Price, then Owner shall promptly pay Contractor the amount theretofore unpaid on Contractor's properly documented invoices submitted for Work performed to the date of termination and Contractor shall have no claim against Owner for any Contractor's Incentive Fee or any other amount.

(d)   Failure of Owner to exercise any of the rights given under this Article shall not excuse Contractor from compliance with the provisions of the Agreement nor prejudice in any way Owner's right to exercise any such rights in respect of such or any subsequent breach by the Contractor.

20.2.   Termination for Owner's Convenience. In addition to Owner's rights to terminate this Agreement as provided in Sections 18.4 and 20.1, Owner shall have the right to terminate this Agreement by giving Contractor written notice, the effective date of such termination being the date of receipt by Contractor of such notice. In the event of such termination Owner shall pay to Contractor: (i) all Contractor's Costs incurred by Contractor as of the effective date of termination, plus Contractor's Fees on that portion of the Work actually performed; (ii) all actual reasonable costs incurred by Contractor in properly terminating the Work including cancellation fees incurred in terminating commitments to third parties, and (iii)the actual cost that is incurred by Contractor in disposition of all completed or uncompleted Work in accordance with Owner's instructions. Owner will then have the obligation within thirty (30) days to advise Contractor as to disposition of any partially completed or uncompleted parts of the Work in the event such items are in Contractor's facilities. The payment terms of this Agreement will apply to payment for termination costs due under this Section 20.2.

20.3.   Termination for Owner's Breach. If insolvency, receivership, or bankruptcy proceedings shall be commenced by or against Owner and not dismissed within thirty (30) days, or if Owner shall make an assignment for the benefit of its creditors, or if Owner shall fail or refuse to perform material obligation hereunder and Owner does not commence to cure such failure or refusal within fifteen (15) days following notice by Contractor, then this Agreement shall be deemed to have been breached by Owner, and

Contractor, without prejudice to and cumulative of all other rights and remedies which Contractor may have as the result of such breach, shall have the option to terminate this Agreement forthwith by giving notice thereof to Owner.

In the event of any such termination, Contractor shall use its best efforts to negotiate termination charges with all subcontractors, vendors, materialmen and suppliers which reflect actual costs incurred in the performance of the Work plus a prorata profit on Work actually performed minus amounts previously paid and any salvage value realized or realizable (excluding consequential damages of the type described in Article XIII).

Upon the settlement of all claims of subcontractors, vendors, materialmen and suppliers and upon presentation of Contractor's final invoice, Owner shall pay to Contractor an amount equal to Contractor's Costs plus Contractor's Fees on Work actually performed minus amounts previously paid and any salvage value realized or realizable (excluding consequential damages of the type described in Article XIII).

ARTICLE XXI

CLAIMS AND LIENS

Except for liens, claims, suits, judgments, and awards related to (i) the Units as a result of Owner's failure to pay the supplier, refurbisher or inspector of the Units, or (ii) breach of this Agreement by the Owner, Contractor agrees to pay, discharge, and hold Owner and its lenders harmless from and against all liens, claims, suits (including counsel fees and other expenses incident thereto), judgment and awards of, or in favor of, subcontractors, vendors, materialmen, suppliers, and laborers which may (whether before or after final payment is made to Contractor hereunder) arise out of the Work. Upon Owner's request, Contractor shall, within five (5) business days, furnish satisfactory proof that all such liens, claims, suits, judgments, and awards have been satisfied or released. If Contractor fails to furnish such satisfactory proof within five (5) business days, Owner may, but is not required to, withhold a sufficient sum from payments then or thereafter due Contractor under Article XI to satisfy such liens, claims, suits (including counsel fees and other expenses thereto) or judgments and awards or may collect such sum directly from Contractor.

ARTICLE XXII

PROPRIETARY INFORMATION

All plans, specifications, designs, documents, and other technical information developed by Contractor or any of its subcontractors, vendors, materialmen and suppliers in connection with the performance of any of the Work shall become the property of Owner at the time they are created and Owner shall have the right to use such materials for any and all purposes in the future. A record copy of all such plans, specifications, designs, documents and other technical information may be retained by Contractor, but Contractor shall obtain Owner's written consent before using such materials on a project for anyone other than Owner or an affiliate of Owner.

All plans, specifications, designs, documents, and other technical information ("Proprietary Information") heretofore or hereafter supplied by Owner with respect to the Work (i) shall remain the property of the Owner, (ii) shall be received and held in confidence by Contractor, and (iii) shall not be used by Contractor for the benefit of itself or others, or reproduced, disclosed to others, published or otherwise used by Contractor except that Contractor may utilize in the performance of the Work such Proprietary Information provided by the Owner.

Notwithstanding anything in this Agreement, expressed or implied, to the contrary, Proprietary Information shall not include any materials or technical information:

- which at the time of the disclosure is generally available to the public or thereafter becomes generally available to the public by publication or otherwise through no act of Contractor; or

- which was in Contractor's possession prior to the time of the disclosure hereunder and was not acquired directly or indirectly from Owner; or

- is independently made available to Contractor as a matter of right by a third party, who is lawfully in possession of such information and not subject to a contractual or fiduciary duty to Owner and who does not require Contractor to refrain from disclosing such information to others.

ARTICLE XXIII

PATENTS AND SIMILAR RIGHTS

Contractor shall, at its sole cost and expense, defend, indemnify and hold harmless the Indemnified Parties from any claim, demand, suit or action which alleges that any part of the Work infringes any patent, proprietary right or copyright, except that Contractor's indemnity does not extend to items manufactured to Owner's design. At Owner's request, Contractor shall assume the defense of any suit or action arising out of any such infringement or claim but Owner shall be entitled to be fully advised and to participate in any such suit or action at Owner's expense (including Owner's attorneys' fees and costs of litigation).

Contractor shall not, without the prior written consent of Owner, settle any infringement or claim or suit based on infringement unless such settlement involves solely the payment of monetary damages, and includes a full and complete release of all Indemnified Parties. Should any claim or suit result in a finding of infringement, Contractor shall, at Contractor's expense, either obtain for Owner a royalty free license to continue to use the patent, proprietary right or copyright found to have been infringed or shall make such changes or replacements as are necessary to render all Work provided pursuant to this Agreement non-infringing.

Contractor shall not enter into any subcontract or other contractual arrangement with any subcontractor, vendor, materialman or supplier which does not contain an indemnity against patent infringement comparable to that stated above.

ARTICLE XXIV

REPRESENTATIONS AND WARRANTIES

24.1. <u>Representations and Warranties by Owner</u>. Owner represents and warrants to Contractor that (a) Owner's execution, delivery and performance of this Agreement has been duly authorized by all necessary corporate action and shall not breach or be in violation of any law, regulation, governmental order, loan agreement, mortgage, or other private contract to which Owner is a party or which is applicable to Owner, (b) Owner is a corporation duly organized and in good standing under the laws of the State of _____, (c) Owner has duly qualified to do business in all states in which the failure to so qualify would have a material adverse effect on Owner's ability to perform its obligations under this Agreement, (d) this Agreement is a valid, legal, and binding obligation of Owner, enforceable against Owner in accordance with its terms, and (e) Owner has or will obtain all licenses, permits and other government approvals necessary in connection with performing its obligations under this Agreement

24.2. <u>Representations and Warranties by Contractor</u>. Contractor represents and warrants to Owner that (a) Contractor's execution, delivery and performance of this Agreement has been duly authorized by all necessary corporate action and shall not breach or be in violation of any law, regulation, governmental order, loan agreement, mortgage, or other private contract to which Contractor is a party or which is applicable to Contractor, (b) Contractor is a corporation, duly organized and in good standing under the laws of the State of _____, (c) Contractor has duly qualified to do business in all states in which the failure to so qualify would have a material adverse effect on Contractor's ability to perform its obligations under this Agreement, and (d) this Agreement is a valid, legal, and binding obligation of Contractor, enforceable against Contractor in accordance with its terms.

ARTICLE XXV

DISPUTE RESOLUTION

25.1   Arbitration. In the event any dispute between Owner and Contractor arising out of or relating to this Agreement cannot be settled amicably by the parties within thirty (30) days after either party has notified the other party of such dispute, then at any time after the expiration of such thirty (30) day period either party may initiate arbitration proceedings by filing a Notice of Demand for Arbitration in writing with the other party and with the American Arbitration Association. Arbitration proceedings shall be in accordance with the then prevailing Construction Industry Arbitration Rules of the American Arbitration Association before a panel of three (3) arbitrators, one (1) of which shall be selected by each of Owner and Contractor and the arbitrators so selected shall select the third arbitrator. A determination by a majority of the panel shall be binding. Reasonable discovery (other than depositions) shall be allowed.

25.2   Performance During Arbitration. The parties shall diligently perform their obligations in accordance with the Agreement pending resolution of the dispute.

25.3   Award. The award rendered by the arbitrators shall be final and binding and judgment may be entered in accordance with applicable law in any court of competent jurisdiction.

25.4   Costs and Expenses. The costs and expenses of the arbitrators and the arbitration shall be allocated among the parties by the arbitrators; provided, however, each party shall in any event bear all of its own costs and expenses (including attorneys' fees).

ARTICLE XXVI

GENERAL MATTERS

26.1.   Non-waiver. Any failure by either party to enforce performance of the terms or conditions of this Agreement shall not constitute a waiver of, or affect its right to avail itself of such remedies relating to any subsequent breach of the terms or conditions of this Agreement.

26.2.   Notices. Unless otherwise provided herein, all notices required or permitted to be given under this Agreement shall be considered duly given if sent by U.S. Certified First Class Mail, postage prepaid, return receipt requested, or by telecopy, addressed as follows, or to such other address as either party may designate in writing:

Owner:

Copy To:

Contractor:

Copy To:

When so transmitted, such notice shall be deemed given as of the date of receipt as shown on the return receipt or the date of sending in the case of a telecopy.

26.3.   Independent Contractor. In the performance of all duties hereunder, Contractor shall be an independent contractor, and Owner shall neither supervise nor control the details of, but rather shall be interested only in the results of, Contractor's services.

26.4.   Amendments. This Agreement shall not be altered, amended or modified unless in a written amendment or Change Order signed by both parties hereto in accordance with Article VIII.

26.5.   Assignment. This Agreement shall not be assigned by either party without the prior written consent of the non-assigning party; provided, however, Owner shall have the right to assign this Agreement without the consent of Contractor to (i) its parent corporation or any affiliate of such parent; or (ii) its lenders as security for loans made or to be made to Owner or (iii) to any entity to which Owner sells the Plant.

26.6.   Governing Law. This Agreement shall be governed by and construed in accordance with the laws of the State of _____, exclusive of any provisions of _____ law which would mandate that the laws of a different jurisdiction should be applied.

26.7.   Headings. Article and Section headings have been inserted in this Agreement for convenience only, are not a part of this Agreement and shall be disregarded when construing this Agreement or any provisions thereof.

26.8.   Entire Agreement. This Agreement and the Exhibits and Annexes hereto constitute the entire understanding of the parties with respect to the subject matter hereof and supersede and negate all prior written or oral undertakings of the parties. The express terms hereof control and supersede any course of performance or usage of trade inconsistent with any of the terms hereof

IN WITNESS WHEREOF, the parties have caused this Agreement to be executed by their duly authorized representatives as of the date first above written.

OWNER: CONTRACTOR:

_____   _____

a _____ corporation   a _____ corporation

By: _____   By: _____

Title: _____   Title: _____

List of Annexes

Annex 1: Legal Description of the Site

Annex 2: Contractor's Key Personnel & Functions

Annex 3: Form of Contractor's Release and Waiver of Liens

Annex 4: Form of Subcontractor's Release and Waiver of Liens

List of Exhibits

Exhibit A: Defined Terms

Exhibit B: Technical Scope Document

Exhibit C: Cash Flow Projections

Exhibit D: Contractor's List of Pre-approved Subcontractors and Vendors

ANNEX 1

PRELIMINARY LEGAL DESCRIPTION OF THE SITE

ANNEX 2

LIST OF CONTRACTOR'S KEY PERSONNEL & FUNCTIONS

Name: Function:

[TO BE PROVIDED BY CONTRACTOR]

ANNEX 3

FORM OF RELEASE AND WAIVER OF LIEN

Know All Men By These Presents, that for valuable consideration paid in full and final payment by _____ ("Owner"), the receipt of which is hereby acknowledged.

_____ ("Contractor") has released and forever discharged, and by these presents does release and forever discharge Owner and its successors, assigns, associates, heirs, executors and administrators, of and from any and every claim, demand, right, lien or right of lien, including but not limited to any right to lien the property described on Exhibit A attached hereto (the "Property"), and cause of action, of any and every kind and nature in law or equity, which the undersigned now has, or may hereafter acquire arising out of the furnishing of labor, materials, service and/or equipment to Owner, or others in connection with the construction, erection and/or alteration of the structure, building or other work or improvements to the Property.

The undersigned warrants and represents that all persons, firms, corporation, partnerships, other entities and all potential lienors which the undersigned has employed or retained, whether directly or indirectly, or which have furnished work, labor, materials, services, equipment, movables and supplies (hereinafter referred to as "Work and Materials") to the undersigned have been paid in full for all Work and Materials furnished as of the date hereof. The undersigned warrants and represents that it has been paid for all Work and Materials furnished to the Property.

Signed, Sealed and Delivered this _____ day of _____, 20_____, before the undersigned Notary Public.

By: _____

Its: _____

_____
Notary Public

EXHIBIT A TO RELEASE AND WAIVER OF LIEN

[Property Description]

ANNEX 4

Form of Partial Waiver of Lien
(By Subcontractors, Suppliers, etc.)

The undersigned, _____, has contracted with or through _____ ("Contractor"), by contract or other agreement dated _____ (the "Agreement") to furnish labor and/or materials and/or services in connection with the construction, erection and/or alteration of the structure, building or other work or improvements to the property more fully described on the attached Exhibit A (the "Property"), owned by ("Owner"), who has entered into a Construction Contract with Contractor, dated _____.

The undersigned hereby acknowledges and confesses that a payment in the amount of $____, being the entire amount due at this time to the undersigned for work completed pursuant to the Agreement and any change orders made in connection therewith, has been received concurrently with the execution hereof. The undersigned certifies and represents that any potential claim which it may have against the Property and/or Owner and/or Contractor has been reduced by the amount of such payment, that all persons, corporations, or other entities who have furnished labor and/or materials to or on behalf of the undersigned pursuant to the Agreement with respect to the work performed on the Property to date ("Completed Work") have been paid, in full, and all such work has been completed, and all such materials have been furnished pursuant to the terms of the Agreement, and that no person, corporation, or other entity has any claim against the undersigned for any labor performed or materials furnished of any kind or character whatsoever in connection therewith.

In connection with the payment referred to above and other good and valuable consideration, the receipt and sufficiency of which are hereby acknowledged and confessed by the undersigned, the undersigned hereby waives and releases all past, present and future liens, claims to liens and lien rights, arising out of or in connection with the Agreement for and on account of labor and materials furnished to, expended upon or used in connection with the Completed Work. The undersigned further agrees to indemnify and to hold the Property, Owner and Contractor, their successors and assigns, harmless from and against any lien, claim or suit of or by any mechanic, supplier, subcontractor or materialman and all costs, expenses and liabilities which the Property, Owner and Contractor may suffer or incur arising out of or connected with the Completed Work.

Signed, Sealed and Delivered this _____ day of _____, 20_____, before the undersigned Notary Public.

[Name of Subcontractor, Supplier, etc.]

By: _____

Its: _____

_____

Notary Public

EXHIBIT A TO PARTIAL WAIVER OF LIEN

[Property Description]

EXHIBIT A

DEFINED TERMS

1.   "Acceptance of Unit Mechanical Completion" shall have the meaning ascribed thereto in Section 4.2.

2.   "Agreement" shall mean the Contract for Engineering, Procurement and Construction Services between Owner and Contractor for the Plant, together with the Annexes and Exhibits thereto.

3.   "Associated Equipment" shall mean that all of the equipment purchased by Owner from _____, except for the four (4) Units and as more particularly described in the TSD.

4.   "Automatic Shutdowns" shall mean those automated devices recommended by the manufacturer or the Contractor that, in their normal operating function, are capable of stopping, by electronic or mechanical means, the operation of other equipment so that such other equipment is not damaged by further operation.

5.   "BOP" shall have the meaning ascribed to such term in the Recitals to this Agreement.

6.   "Change Order" shall have the meaning scribed thereto in Section 8.1(c).

7.   "Commencement Date" shall have the meaning ascribed to such term in Section 4.1.

8.   "Contractor" shall have the meaning ascribed to such term in the Preamble to the Agreement.

9.   "Contractor's Costs" shall mean costs reasonably and necessarily incurred in the proper performance of the Work and paid or to be paid to Contractor. Such costs shall be at rates not higher than the standard rates paid in the locality where the Work is performed, except with the consent

of Owner, and shall include the items set forth below in items (i) through (xvi).

(i)   compensation paid for labor, including but not limited to engineering, design and construction labor, in the direct employ of Contractor in the performance of the Work, including such insurance and other benefits (exclusive of special bonuses or other special payment not a part of Contractor's normal compensation program), if any, as may be payable with respect thereto;

(ii)   compensation paid to employees of Contractor when stationed at the field office in whatever capacity employed, it being specifically understood and agreed that employees of Contractor, including the general manager and the Project Manager of Contractor, engaged in managing, administering or accounting for the Work or in expediting the production or transportation of materials or equipment to be incorporated into the Work shall be considered as stationed at the field office and their compensation paid for that portion of their time spent on the Work;

(iii)   costs of contributions, assessments or taxes for such items as unemployment compensation and social security insofar as such cost is based on compensation or other remuneration paid to employees of Contractor included in Contractor's Costs;

(iv)   cost of coach class transportation, and other reasonable traveling, meal and hotel expenses of Contractor's employees incurred in discharging duties connected with the Work;

(v)   cost of all materials, supplies and equipment incorporated into the Work, including reasonable costs of transportation thereof (with due credit to be given at the completion of the project for surplus materials and supplies);

(vi)   expenses incurred pursuant to this Agreement by Contractor in respect to subcontractors, materialmen and/or other parties retained by, through or under Contractor for Work required to be performed to design or construct the facility described in the TSD;

(vii) cost, including transportation and maintenance, of all materials, including all consumables and expendables used in the engineering and design of the facility described in the TSD, and all supplies, equipment, temporary facilities and hand tools which are not owned by workmen and which are consumed in the performance of the Work, and cost less salvage value in respect to such items which are used but not consumed and which remain the property of the Contractor (no computers are to be charged to Owner pursuant to this provision);

(viii) rental charges of all necessary machinery and equipment, exclusive of hand tools, used at the site of the Work, whether rented from Contractor or others, including installation, minor repairs and replacements, dismantling, removal, transportation and delivery costs thereof, at rental charges consistent with those prevailing in the area of the Work;

(ix) cost of premiums for all bonds, guarantees, warranties and insurance which Contractor is required to purchase and/or maintain in order to carry out the Work;

(x) sales, use or similar taxes related to the Work and for which Contractor is liable imposed by any governmental authority;

(xi) permit fees, royalties and deposits lost for causes other than the negligence of Contractor;

(xii) expenses of reconstruction and/or other costs to replace and/or repair damaged materials and supplies provided that Contractor is not compensated for such expenses and/or costs by insurance otherwise and that such expenses and/or costs have resulted from causes other than the fault or omission of Contractor, it being specifically understood and agreed that such expenses and/or costs shall include costs incurred as a result of any act or neglect of Owner or of any of Owner's representatives or of any separate contractor of Owner;

(xiii)   expenses for such items as shipping, delivery, facsimile transmissions, telegrams, long distance communications, telephone service at the Site, postage, expressage, copying, computer usage (with computer usuage to billed at $____/hr of usuage for engineering applications and no charge is to be made for usuage for word processing) and petty cash items purchased in connection with the Work;

(xiv)   cost of removal of debris;

(xv)   cost incurred due to an emergency affecting the safety of persons and property; and

(xvi)   other costs incurred in the performance of the Work if and to the extent such costs are approved in writing by Owner. Except as otherwise expressly provided in this Agreement, the term "Contractor's Costs" shall not include any of the items set forth below in items (zi) through (ziv):

(zi)   salaries or other compensation of marketing personnel, general mangers, auditors or clerical personnel of Contractor at the principal office or branch offices of Contractor, unless temporarily assigned for a continuous period of at least fourteen (14) days to performance of the Work;

(zii)   expenses of the principal or branch offices of Contractor other than the field office at the Site;

(ziii) any part of the capital expenses of Contractor, including interest on capital of Contractor employed for the Work; and

(ziv) general administrative or other indirect overhead expenses of any kind, except as may be expressly included above in items (i) through (xvi).

10. "Contractor's Fees" shall mean fifteen and one-half percent (15.5%) of Contractor's Costs, which represents the payment to Contractor for its overhead and general and administrative expenses, other costs and expenses not reimbursable by Owner as Contractor's Costs, and Contractor's profit.

11. "Contractor's Incentive Fee" shall mean fifty percent (50%) of the difference between (a) the sum of (i) Contractor's Costs and (ii) Contractor's Fees, and (b) the Maximum Contract Price minus the sum of (iii) Liquidated Damages, if any, and (iv) any amount reasonably determined by Owner to be necessary to correct any portion of the Work not performed in accordance with the Agreement.

12. "Deficiency Notice" shall have the meaning ascribed to such term in section 4.2.

13. "Effective Date" shall have the meaning ascribed to such term in the Preamble to the Agreement.

14. "Final Acceptance" shall have the meaning ascribed to such term in Section 4.5.

15. "Fully Loaded Conditions" shall mean the performance conditions (flow rates, temperatures and pressures) required of the BOP equipment, as specified by Contractor supplied and Owner approved computer operating simulations of the Units and BOP.

16. "Guaranteed Plant Completion Date" shall mean the date which is (i) _____ days after the issuance of the Notice to Proceed with Construction or (ii) _____ days after the last of the turbine or the generator for each Unit is available for shipment from the point of refurbishment, whichever occurs later.

17. "Guaranteed Unit Mechanical Completion Date" shall mean for each of the four (4) Units, the date which is (i) _____ days after the issuance of the Notice to Proceed with Construction or (ii) _____ days after the last of the turbine or the generator for each Unit is available for shipment from the point of refurbishment, whichever occurs later.

18.   "Liquidated Damages" shall have the meaning ascribed to such term in Section 10.1.

19.   "Maximum Contract Price" shall have the meaning ascribed to such term in Section 5.1.

20.   "Milestone Dates" shall mean the Guaranteed Unit Mechanical Completion Date, and the Guaranteed Plant Completion Date.

21.   "Notice to Proceed with Construction Commencement" shall have the meaning ascribed to such term in Section 4.1.

22.   "Owner" shall have the meaning ascribed to such term in the preamble to this Agreement.

23.   "OSHA" shall have the meaning ascribed to such term in Section 12.1.

24.   "Performance Testing Period" shall have the meaning ascribed to such term in Section 4.4.

25.   "Plant" shall have the meaning ascribed to such term in the Recitals to this Agreement.

26.   "Plant Commercial Operation" shall have the meaning ascribed to such term in Section 4.4.

27.   "Plant Completion Date" shall have the meaning ascribed to such term in Section 4.5.

28.   "Project Manager" shall have the meaning ascribed to such term in Section 2.2.

29.   "Project Schedule" shall mean the detailed critical path schedule to be developed by Contractor and approved by Owner which shall include, without limitation, the milestones set forth in the TSD.

30.   "Proprietary Information" shall have the meaning ascribed to such term in Article XXII.

31.   "Punch List" shall mean a written list to be jointly prepared by Owner and Contractor when the Work, or a designated portion thereof, is Substantially Completed, which list shall specify items to be completed or corrected by Contractor.

32.   "Rain Day" or "Rain Days" shall mean any day or part of a day on which performance of any of the Work necessary to complete the Plant to the point of Unit Mechanical Completion is prohibited or significantly impaired due to precipitation, lightning or wind or the effects of precipitation, lightning or wind at the location where such portion of the Work is being performed.

33.   "Retainage" shall have the meaning ascribed to such term in Section 6.1(b).

34.   "Site" shall have the meaning ascribed to such term in the Recitals to this Agreement.

35.   "Subcontract" shall have the meaning ascribed to such term in Section 3.1.

36.   "Substantial Completion" or "Substantially Completed" shall mean that all of the Work has been completed in accordance with the terms of this Agreement, including the acceptance of Turnover Documents by Owner, to the point where Owner, subject to and upon performance of any separate work to be done by Owner or any of Owner's other contractors and subject to Unit Start-up, can take possession of the applicable Unit and to commence operations with that Unit with Minimal Interference.

37.   "TSD" shall mean the Technical Scope Document prepared by Contractor and incorporated into the Agreement as Exhibit B.

38.   "Turnover Documents" shall mean manuals prepared by Contractor for each Plant system (as described in the TSD) detailing the testing procedures to confirm readiness for Unit Mechanical Acceptance

and turnover of control of the Plant systems (as described in the TSD) to Owner.

39.   "Unit" and "Units" shall have the meaning ascribed to such term in Section 1.2(d).

40.   "Unit Initial Commercial Operation" shall have the meaning ascribed to such term in Section 4.3(b).

41.   "Unit Mechanical Completion" shall have the meaning ascribed to such term in Section 4.2. "Mechanically Complete" shall have the correlative meaning.

42.   "Unit Mechanical Completion Notice" shall have the meaning ascribed to such term in Section 4.2.

43.   "Unit Start-up" shall have the meaning ascribed to such term in Section 4.3.

44.   "Work" shall have the meaning ascribed to such term in Section 1.2(a).

Courtesy of Trent A. Gudgel, Hall, Estill, Hardwick, Gable, Golden & Nelson PC

APPENDIX B

POWER PLANT CONSTRUCTION CONTRACT

THIS CONSTRUCTION CONTRACT is entered into and effective as of the _____ day of _____, 20___, by and between _____ (herein called "Contractor"), and _____ (herein called "Owner").

WHEREAS, Contractor is experienced in the design and construction of the type of facility desired by Owner and has submitted to Owner a certain proposal to perform the design and construction work for the facility desired by Owner, all as more fully set forth herein.

NOW, THEREFORE, in consideration of their mutual covenants and other good and valuable consideration, the receipt and sufficiency of which are hereby acknowledged, the parties agree as follows:

1.   Definitions. In this Construction Contract all defined terms are set forth in Appendix 1and shall have the meaning herein as stated in Appendix 1.

2.   Contract and Exhibits.   The following documents shall comprise the Contract and shall be referred to as the "Contract:"

A.   This Construction Contract between _____ and _____ Dated _____.

B.   Appendix 1: Contract Term Definitions

C.   Exhibit A: The Technical Scope Document

D.   Exhibit B: Partial Waiver of Liens and Indemnity Agreement

E.   Exhibit C: Final Waiver of Liens and Indemnity Agreement

F.   Exhibit D: _____

The above Exhibits are attached hereto and incorporated herein by reference. If there is any conflict between the terms of this Construction Contract and the Exhibits, then the provisions of this Construction Contract shall prevail.

3.   Scope of Work. Contractor agrees to perform the Work and shall furnish all labor, material, supplies, equipment and services necessary to complete the Work in a good and workman like manner consistent with the TSD. Owner will provide the Site, air permit and natural gas pipeline to supply the Facility.

4.   Condition Precedent and Owner Representations Regarding Certain Equipment.

A.   It is a condition precedent to Contractor's obligation to perform the Work that Owner shall, without cost or expense to Contractor, procure and provide all materials, equipment, labor and services necessary for the procurement and delivery of the Units, _____, and all other Owner-supplied equipment and materials as specified in the TSD (collectively referred to as "Owner's Equipment").

B.   The Units shall be transported to the Site by _____ (date). Owner shall furnish Contractor a written report on a daily basis during the time when the Units are being transported to the Site that updates Contractor on the estimated time of arrival to the Site.

5.   Contract Price. The price to be paid by Owner and accepted by Contractor for the performance of the Work under this Contract is _____ Dollars ($_____) (referred to herein as the "Contract Price"). The Contract Price is fixed and shall not be increased for any reason, except as otherwise provided under the terms of this Contract or by the agreement of the parties. It is specifically understood and agreed that Owner shall pay to Contractor for the performance of the Work the Contract Price as adjusted under the terms of this Contract.

6.   Completion of Work, Liquidated Damages and Incentive.

A.   Contractor has prior to the date of this Contract proceeded with the procurement of certain equipment and materials and certain engineering and design services required to build the Facility, and Contractor shall complete the engineering and design services required to complete the Work with due diligence, due regard being given to interruptions resulting from any cause Beyond the Reasonable Control of Contractor. If Contractor receives a notice to proceed with the Work on or before _____, 20__, Contractor agrees to complete the Work so as to meet the following Milestone Dates, subject to extensions of the dates as allowed by this Contract:

[Milestone Dates]

The construction schedule for the Work includes _____ Weather Days through _____. Additional Weather Days shall extend the construction schedule for all Milestone Dates on a day-for-day bases up to _____ days. If the notice to proceed is issued by Owner prior to _____, 20___, each day between and including the date of the notice to proceed and _____, 20___ shall be added to the construction schedule and considered as Weather Days in addition to the _____ Weather Days included in the original construction schedule.

B.   Contractor acknowledges that, in the event that the Work has not progressed to a point to allow the Units to generate power to the substation at the rated capacity on or before _____, 20___, as that date may be extended under the terms of this Contract, Owner will incur substantial delay damages, including without limitation lost profits, and Owner and Contractor acknowledge that the actual amount of any such delay damages incurred by Owner will be difficult or impossible to calculate with accuracy. In the event that the Work has not progressed to a point to allow the Units to generate power to the substation at the rated capacity on _____, 20___, as that date may be extended under the terms of this Contract, Owner may, at the time of final payment, offset against final payment, as liquidated damages an amount equal to _____ Dollars ($_____) per Unit per day for each day after _____, 20___ that the Work has not progressed to a point to allow the particular Unit to generate power to the substation at the rated capacity. On the date that a Unit is capable of generating power to the substation at the rated capacity, Owner shall not be entitled to recover any more liquidated damages relating to that Unit. It is specifically understood and agreed by Owner and Contractor that the liquidated damages shall be in lieu of, and shall be deemed to be full compensation by Contractor to Owner for and in respect to any delay damages actually incurred by Owner. The maximum amount of liquidated damages Contractor shall be obligated to pay to Owner under this Contract shall not exceed $_____.

C.   In the event Contractor achieves Balance of Plant Operational prior to _____, 20___, then Contractor shall receive as an incentive of _____ Dollars ($_____) per day for each day prior to _____, 20___ that Balance of Plant Operational is achieved. Such incentive shall be an addition to the Contract Price, and shall be paid to Contractor as part of the final payment in accordance with paragraph 8A.

D.   In the event that the performance or the completion of the Work is delayed at any time by any act or omission of Owner, including but not limited to failure to timely procure and deliver Owner's Equipment, or of any employee, agent or representative of Owner, or of any separate contractor employed by Owner, by changes or alterations in the Work not caused by any fault or omission of Contractor, by strikes or other labor disputes, lockouts, fire, embargos, windstorm, flood, earthquake or other acts of God, acts of war, changes in public laws, regulations or ordinances enacted after the date of execution of this Contract, by acts of public officials not caused by any fault or omission of Contractor, by an inability to obtain materials or equipment not caused by any fault or omission of Contractor, or by any other cause Beyond the Reasonable Control of Contractor and as a result of which the critical path then employed by Contractor in scheduling the design and construction of the Work cannot be reasonably revised to accommodate such delay without additional cost to Contractor and without affecting the Milestone Dates, the Milestone Dates shall be extended for a reasonable period as a consequence of such delay. Any claim for an extension of the Milestone Dates or of the time within which to substantially complete the Work required to be performed by Contractor shall be made in writing to Owner not later than fifteen (15) days after Contractor has notice of the cause for the delay upon which claim for extension is based. In the event of a continuing delay, only one claim for extension shall be required to be made. Contractor, upon making a claim for extension, shall provide Owner with an estimate of the probable effect, which the cause of delay will have upon the progress of the Work.

7.   Terms of Payment.

A.   Owner shall be obligated to pay the Contract Price to Contractor in accordance with the terms of this Contract. During the performance of the Work, Contractor shall submit to Owner on or about the tenth day of each month an application for progress payment covering all labor, materials and other work furnished by Contractor and/or invoiced to Contractor, by any subcontractor, laborer, supplier or materialman hired by Contractor, during the preceding month, including a portion of Contractor's Fee earned as a result of the performance of such Work. Subject to the terms set forth in this paragraph, _____ percent (_____%) of the amount attributable to the Work and _____ percent (_____%) of Contractor's Fee included in any application for progress payment shall be paid to Contractor within fifteen (15) days after the receipt of any application for progress payment by Owner. The unpaid balance of any application for progress payment will be retained until final payment per paragraph 8A.

B.   Contractor shall submit to Owner with each application for progress payment a monthly job cost report showing an itemized listing of the Cost of the Work incurred by Contractor for labor, materials and other work included in such application for payment submitted. Contractor shall submit to Owner satisfactory evidence reasonably requested by Owner reflecting the proper payment by Contractor of all indebtedness incurred for labor, materials and other work, the costs of which have been included in any application for progress payment submitted and paid by Owner. Contractor shall also provide to Owner, as and when received by Contractor, (i) unconditional releases and waivers of lien issued by all parties retained by Contractor in connection with the Work who have rendered services, performed work or furnished materials, the cost of which is included in any application for payment, under this Contract and is in excess of Five Thousand Dollars ($5,000.00), evidencing that such parties have been paid all amounts, less any applicable retainage, attributable to materials, labor and/or other work, the costs of which have been included in any application for payment submitted and paid by Owner; (ii) conditional or unconditional releases and waivers of lien, as appropriate, issued by all parties retained by Contractor in connection with the Work who have rendered services, performed work or furnished materials, the costs of which are included in any application for payment, under this Contract and is in excess of Five Thousand Dollars ($5,000.00) indicating any amounts, including applicable retainage, that are due or owing to such parties for materials, labor and/or other work, the costs of which have been included in any application for payment submitted and not paid by Owner; and (iii) a waiver of liens and indemnity agreement issued by Contractor in form and substance substantially identical to the partial waiver of liens and indemnity agreement attached as Exhibit B.

C.   All applications for payment under this Contract, including the application for final payment, shall be subject to the approval of Owner. Owner shall cause all applications for payment under this Contract, including the application for final payment, to be approved by Owner immediately upon receipt by Owner, unless Owner shall in good faith dispute any amount included in an application for payment or unless an application for payment shall not be reasonably in accordance with the terms of this Contract. Owner shall not be entitled to withhold any approval of any application for payment under this Contract if such application for payment is reasonably in accordance with the terms of this Contract and payment of the amount shown in such application for payment is due to Contractor pursuant to the terms and provisions of this Contract.

D.   Owner may withhold any progress payment to be made to Contractor in whole or in part and/or may, because of subsequently discovered evidence or subsequent inspection, nullify all or any part of any progress payment previously made to Contractor to the extent reasonably necessary to protect Owner from loss because of:

    1.   defective work and/or work not performed by Contractor in accordance with the TSD, provided that Contractor has been given written notice by Owner of such defective and/or nonconforming work and has failed to remedy such defective and/or nonconforming work promptly after being given written notice by Owner of such defective and/or nonconforming work;

2. claims filed or reasonable evidence indicating the probable filing of such claims by third parties as a result of any act or omission of Contractor; or

3. failure of Contractor to make payments properly to subcontractors, laborers and/or materialmen retained by Contractor.

All actions which Owner is entitled to take with respect to the withholding and/or nullifying of all or any part of any progress payment under this Contract shall be reasonable and Owner shall promptly notify Contractor in writing of any and all grounds for the withholding and/or nullifying of any progress payment. When all or any part of any progress payment shall be withheld, and the grounds for such withholding are removed, Owner shall promptly pay Contractor all or such part of such progress payment so withheld because of such grounds. Notwithstanding the foregoing, if Contractor fails to make any payment to a subcontractor, laborer and/or materialmen retained by Contractor, Owner shall not be entitled to withhold any progress payment if and to the extent Contractor furnishes Owner with a surety bond in a form and in an amount that complies with all applicable statutory requirements to protect Owner from loss.

E. If Owner should in good faith dispute any amount included in any application for progress payment submitted by Contractor, Owner shall within fifteen (15) days after receipt of such application for payment, notify Contractor in writing of the general nature of such dispute and pay in accordance with this Contract any amount included in such application for payment and not so disputed. No amounts included in any application for payment shall be considered to be disputed in good faith by Owner unless Owner has notified Contractor of the dispute as set forth above and has paid in accordance with the terms of this Contract all amounts included in any application for payment that Owner does not dispute.

F. Contractor shall continue to proceed with the Work and shall maintain the progress of the Work in the event that Owner shall dispute in good faith any amount included in any application for payment provided that Owner shall, notwithstanding such dispute, continue to perform in accordance with the terms of this Contract, all obligations of Owner not disputed.

G.   Notwithstanding anything to the contrary set forth herein, Contractor shall not be required as a condition precedent to the obligation of Owner to make any payment, including any portion of the final payment, to provide to Owner any releases and/or waivers of lien issued by any party retained by, through or under Contractor in connection with the Work (i) if Contractor furnishes Owner with documentation clearly evidencing that such party does not have any right at law to record or file any mechanic's lien affecting the Work or (ii) if Contractor furnishes Owner with a surety bond in a form and in an amount that complies with all applicable statutory requirements to protect Owner from loss because of the recording or filing, or potential recording or filing, by such party of any mechanic's lien in respect to the Work.

8.   <u>Final Payment</u>.

A.   Contractor shall submit to Owner a final application for payment immediately after Substantial Completion of the Work. Final payment is due and payable by Owner within sixty (60) days after Substantial Completion of the Work in accordance with the Contract, unless Contractor has failed to Fully Complete the Work within such sixty (60) day period, in which case final payment shall be due and payable by Owner in two installments, the first installment being equal to such final payment less an amount equivalent to twice the value of any uncompleted Work and being due and payable by Owner upon the expiration of such sixty (60) day period and the second installment being equal to the balance of such final payment and being due and payable promptly upon Full Completion of the Work.

B.   Contractor shall certify and submit to Owner documentation reasonably requested by Owner relating to: (a) the Cost of the Work incurred by Contractor for labor, materials and other work included in the application for final payment, and (b) the calculation of any savings as set forth in paragraph 9. Further, Contractor shall make available to Owner at Contractor's principal place of business satisfactory evidence reasonably requested by Owner, including without limitation, copies of bills, purchase orders, contracts, invoices, receipts, releases, canceled checks or payroll records, reflecting the proper payment by Contractor of all indebtedness incurred for materials, labor and other work, the cost of which has been included in any application for payment submitted and paid by Owner. Further, Contractor shall provide to Owner (i) unconditional releases and waivers of lien issued by all parties retained by Contractor in connection with the Work who have rendered services, performed work or furnished materials, the cost of which is included in any application for payment under this Contract showing that such parties have been paid all amounts, less any applicable retainage, attributable to materials, labor or other work, the cost of which has been included in any application for payment submitted and paid by Owner; (ii) conditional or unconditional releases and waivers of lien, as appropriate, issued by all parties retained by Contractor in connection with the Work who have rendered services, performed work or furnished materials, the cost of which is included in any application for payment under this Contract indicating any amounts, including applicable retainage, that are due or owing to such parties for materials, labor or other work, the cost of which has been included in any application for payment submitted and not paid by Owner; and (iii) a waiver of liens issued by Contractor in a form and substance substantially identical to the final waiver of liens and indemnity agreement attached as Exhibit C.

C.   If Owner shall in good faith dispute any amount included in the final application for payment submitted by Contractor, Owner shall notify Contractor in writing within thirty (30) days after receipt of such application for payment of the general nature of such dispute and pay in accordance with this Contract any amount included in the final application for payment and not so disputed, it being specifically understood that no amounts included in the final application for payment shall be considered to be disputed in good faith by Owner unless Owner has so notified Contractor of such dispute as set forth above and has paid in accordance with this Contract all amounts included in the final application for payment and not disputed.

9.   Savings. In the event that the sum of the Cost of the Work and Contractor's Fee is less than the Contract Price, as finally adjusted, one hundred percent (100%) of the difference shall be deemed to be a savings. Any savings shall be determined at the time of final payment and shall

decrease the Contract Price payable by Owner to Contractor in an amount equal to _____ percent (_____%) of the total dollar value of the savings, subject to adjustment as provided in this paragraph 9. The percentage of savings to Owner shall be reduced by one percent (1%) per day, up to a maximum reduction in savings to Owner of fifteen percent (15%), for each day of delay in the performance of the Work caused by any instruction, act or omission of Owner, its employees, agents, representatives, or any other contractor, materialman, or laborer hired by Owner to perform work or supply materials for the construction of the Facility, that prevents the Facility from being capable of generating power to the substation at the rated capacity by _____. The percentage of savings to Owner shall also be reduced by one percent (1%) per day, up to a maximum reduction in savings to Owner of fifteen percent (15%), for each day prior to _____, as that date may be extended under the terms of this Contract, that the Facility is capable of generating power to the substation at the rated capacity. If Contractor does not complete the Work so that the Facility is capable of generating power to the substation at the rated capacity by _____, as that date may be extended by this Contract, and if Contractor's failure to so complete the Work is in no way caused by any instruction, act or omission of Owner, its employees, agents, representatives, or any other contractor, materialman or laborer hired by Owner, the Contract Price payable by Owner to Contractor shall be reduced by one hundred percent (100%) of the total dollar value of the savings.

10. <u>Construction of the Work</u>. No payment hereunder by Owner and no provision in the TSD shall relieve Contractor of responsibility for faulty materials or workmanship incorporated into the Work. Contractor warrants that all Work done under this Contract shall be without defects, including without limitation defects in design, and free of faulty materials and workmanship and agrees, immediately upon receiving notification from Owner, to remedy, repair or replace all defective and/or faulty Work which may appear at any time, or from time to time, during a period beginning with commencement of construction of the Work and ending one year after the date of full completion of the Work. The foregoing warranty of Contractor applies to the remedy, repair or replacement of defects or imperfections which may appear as a result of faulty designs prepared by Contractor and/or by any party retained by, through or under Contractor in

connection with the design of the Work, but the foregoing warranty of Contractor does not guarantee against damage to the Work sustained by lack of normal maintenance or as a result of changes or additions to the Work made or done by persons not directly responsible to Contractor, except where such changes or additions are made in accordance with Contractor's directions. No guarantee furnished by a party other than Contractor with respect to equipment manufactured or supplied by such party shall relieve Contractor from the foregoing warranty obligation of Contractor. The warranty period shall not apply to latent defects, and with respect to such defects, the applicable statutes of limitations and repose shall apply. **Contractor makes no guarantee or warranty as to the quality, fitness or merchantability of the Work, express or implied, except as set forth in this paragraph 10.**

11.   Permits and Regulations.

A.   Any permits, other than the building permit, licenses and authorizations necessary for the construction of the Work and required to be obtained from any governmental authority shall be secured and paid for by Owner. It shall be the responsibility of Contractor to obtain the building permit and to make certain that the Work complies with all applicable building codes and all bulk and dimensional requirements set forth in any applicable zoning ordinances; provided, however, that it shall be the responsibility of Owner, and not of Contractor, to make certain that the facility described in the TSD will be allowed to be used for the purpose intended in the zoning district in which the facility is situated.

B.   Contractor shall give all notices and comply with all applicable laws, ordinances, rules, regulations and orders of any public authority bearing on the performance of the Work. Without in any way limiting the responsibilities of Contractor set forth above, if the Work or the TSD are at variance in any respect with any such laws, ordinances, rules, regulations or orders, Contractor shall promptly notify Owner in writing and shall promptly modify the Work and/or the TSD so that the Work and the TSD will be in compliance with such laws, ordinances, rules, regulations and orders, it being specifically understood (a) that the Contract Price, and the Milestone Dates shall not be adjusted for or in respect to any such variances unless the Work and/or the TSD are required to be modified as a result of a change in such laws, ordinances, rules, regulations or orders occurring after execution of this Contract and (b) that if the Work and/or the TSD are required to be modified as a result of a change in such laws, ordinances, rules, regulations or orders occurring after execution of this Contract, the Contract Price and the Milestone Dates shall be equitably adjusted in accordance with paragraph 16 upon claim made by Contractor to Owner as set forth in paragraph 16. If Contractor performs any Work contrary to such laws, ordinances, rules, regulations or orders Contractor shall assume full responsibility therefor and shall bear all costs attributable thereto.

12.   Protection of Persons and Property.

A.   Contractor shall be responsible for initiating, maintaining, and supervising all safety precautions and programs in connection with the construction of the Work. Contractor shall take reasonable precautions for the safety of, and shall provide reasonable protection to prevent damage, injury or loss to:

1. all employees on the Work and all other persons who may be affected thereby;

2. all Work and all materials and equipment to be incorporated therein, whether in storage on or off the Site, under the care, custody, or control of Contractor; and

3. other property at the Site or adjacent thereto, including trees, shrubs, lawns, walks, pavements, roadways, structures and utilities not designated for removal, relocation or replacement in the course of construction.

Contractor shall at all times enforce strict discipline and good order among those performing the Work and shall not intentionally permit any unfit person or anyone not skilled in the task assigned to be employed or remain employed on the Work.

B.   In an emergency affecting the safety of persons or property, Contractor shall act at its discretion to prevent threatened damage, injury or loss. Any additional compensation or extension of time claimed by

Contractor on account of any such emergency shall be determined by mutual agreement between Owner and Contractor; provided, however, that Contractor shall not be entitled to any additional compensation or any extension of time caused by any fault or omission of Contractor or any party for whose acts or omissions Contractor is responsible.

13.   Inspection of Work.

A.   Owner and Owner's representatives shall at all times have access to the Work wherever it is in preparation or progress, and Contractor shall provide proper facilities for such access and for inspection of the Work.

B.   Contractor shall, as part of the Work, be responsible for the procurement and coordination of all testing and inspection of the Work required by any governmental authorities having jurisdiction over the Work and/or required to insure that the Work is being constructed without defects and in accordance with the TSD. Any required certificates and/or reports pertaining to such testing or inspection shall be secured by Contractor and promptly delivered by Contractor to Owner. Contractor shall advise Owner in advance of the dates, times and locations of all such testing or inspection so that Owner may, if Owner desires, observe such testing or inspection.

14.   Supervision and Review of Drawings.

A.   Contractor shall keep on the Work during its progress a competent project manager and any necessary assistants. Such project manager shall represent Contractor and all directions given to such project manager shall be as binding as if given to Contractor. Important directions by Owner to Contractor shall be confirmed in writing to Contractor. Other directions by Owner to Contractor shall be so confirmed upon the request of Contractor in each case.

B.   As promptly as reasonably possible after the execution of this Contract and consistent with the construction schedule, Contractor shall cause to be prepared and submitted to Owner for approval conceptual project drawings as described in the TSD and the project procedures manual dated _____. Owner shall promptly notify Contractor in writing if Owner, after reviewing any drawing or diagram submitted by Contractor to Owner in respect to any specific aspect of the Facility, approves such drawing or diagram or has reasonable objection to and does not approve such drawing or diagram. If Owner fails so to notify Contractor in writing within ten (10) days after the submission by Contractor of any such drawing or diagram, the drawing and diagram will be considered approved without objection by Owner. However, this does not relieve Contractor from the responsibility of compliance with acceptable engineering and design standards. Contractor shall promptly cause any such drawing or diagram to which Owner has reasonably objected as aforesaid to be changed or revised and to be submitted again to Owner for approval in accordance with the procedure set forth above.

15.   <u>Changes in the Work</u>.

A.   Owner, without invalidating this Contract, may order extra Work or make changes by altering, adding to, or deducting from the Work, provided that the Contract Price, Contractor's Fee and Milestone Dates shall be adjusted accordingly. Unless otherwise specified, all Work shall be executed under the conditions of this Contract and any claim for extensions of time caused thereby shall be adjusted at the time of ordering such change.

B.   Any change in the Contract Price resulting from a change in the Work shall be determined in one or more of the following ways:

1. by mutual acceptance of a lump sum itemized and supported by substantiating data;
2. by unit prices agreed upon by the parties;
3. by cost to be determined in a manner agreed upon by Owner and Contractor and by an increase or decrease in the Contract Price based upon such cost and an adjustment in Contractor's Fee consistent with such change in the Contract Price; or
4. by the method provided in paragraph 1 5C.

C.   If none of the first three methods set forth above is agreed upon, Contractor, provided Contractor receives a written order signed by Owner, shall promptly proceed to cause the requested change in the Work to be performed. The change in the Contract Price shall then be determined on the basis of the sum of any increase and/or decrease in the Cost of the Work attributable to the change and any increase and/or decrease in Contractor's Fee attributable to the change. In such case, Contractor shall keep and present to Owner an itemized statement, together with appropriate supporting data, of the Cost of the Work attributable to the change. Pending final determination of the change in the Contract Price, payments on account of the change shall be made based upon an application for payment covering actual expenditures of Contractor attributable to the change. The amount of credit to be allowed by Contractor to Owner for any deletion or change which results in a net decrease in the Contract Price shall be determined on the basis of a reasonable estimate by the Contractor of the Cost of the Work deleted or changed. When both additions and credits covering related Work or substitutions are involved in any one change which results in a net increase in the Contract Price, the change in the Contract Price shall include an increase in Contractor's Fee calculated only upon the basis of the net increase in the Cost of the Work attributable to the change.

D.   The Contract Price, Contractor's Fee and Milestone Dates shall be equitably adjusted in accordance with paragraph 16 should Contractor incur costs of reconstruction and/or costs to replace and/or repair damaged materials and supplies, provided that Contractor is not compensated for such expenses and/or costs by insurance or otherwise and that such costs have resulted from causes other than the fault or neglect of Contractor or of any employee, agent or representative of Contractor. In the event that any such reconstruction, replacement and/or repair is required and Contractor is placed in charge thereof, Contractor shall be paid for said construction, replacement and/or repair services a Contractor's Fee consistent with that set forth in Appendix 1, definition E, said Contractor's Fee on account of said reconstruction, replacement and/or repair services being an addition to the Contract Price.

16.   Claims and Extensions. If Contractor claims that any instruction provided by Owner and/or any act or omission, whether intentional or unintentional, of Owner or of any employee, agent or representative of Owner, involves extra cost and/or that, in accordance with paragraph 1 1B, any change has occurred after execution of this Contract in applicable laws,

ordinances, rules, regulations or orders of any public authority having jurisdiction over the Work and/or that, in accordance with paragraph 1 5D, the Contract Price should be increased as a result of Contractor having incurred expenses of reconstruction and/or costs to replace and/or repair damaged materials and supplies, and/or that, in accordance with paragraph 21, Concealed and/or Unknown conditions have been encountered, Contractor shall give written notice and explanation thereof within a reasonable time after the receipt of such instructions, the occurrence of such act or omission, the learning about the effect of such change and/or the incurring of such expenses and/or costs and, provided that such claim is reasonable and supported by clear evidence, such claim shall be treated as a change in the Work for which the Contract Price, Contractor's Fee and Milestone Dates shall be equitably adjusted. Notwithstanding anything to the contrary set forth herein, the Contract Price and/or Contractor's Fee shall not be increased, and the Milestone Dates shall not be extended, if and to the extent that such increase or extension shall be attributable exclusively to the fault or omission of Contractor and/or any party retained by, through or under Contractor in connection with the Work.

17.   Separate Contracts.

A.   Owner reserves the right to let other contracts in connection with the construction of portions of the Work which are not being performed hereunder by Contractor. Contractor shall afford other contractors reasonable opportunity for the introduction and storage of their materials and the execution of their work and shall properly connect and coordinate the Work with the work of such other contractors.

B.   If the proper execution of any part of the Work depends upon the work of any such other contractor, Contractor shall inspect and promptly report to Owner any apparent discrepancies in such work that render it unsuitable for such proper execution and results. Failure of Contractor so to inspect and report shall constitute an acceptance of such other contractor's work as fit and proper to receive the Work, except as to latent defects and defects which may develop in such work after the execution of such work.

18.   <u>Use of Premises</u>. Contractor shall confine operations at the Site to areas permitted by law, ordinances, permits and the TSD, and shall prevent the Site from being unreasonably encumbered with any materials or equipment. Contractor shall not permit any part of the Work to be loaded with a weight so as to endanger the safety of persons or property at the Site.

19.   <u>Cleaning Up</u>. Contractor shall at all times keep the Site free from accumulations of waste material or rubbish caused by the performance of the Work by Contractor, and at the completion of construction of the Work, Contractor shall remove from the Site all rubbish and all tools, scaffolding and surplus materials caused by, used in or resulting from the Work and shall leave the Work "broom clean," or its equivalent, unless more exactly specified hereunder.

20.   <u>Site Representations</u>.

A.   Owner warrants and represents that Owner has, and will continue to retain at all times during the course of construction of the Work, record legal title to the land upon which the Facility is to be constructed and that such land is properly zoned so as to permit the construction and use of the Facility. Owner further warrants and represents that title to such land is free of any easements, conditions, limitations, special permits, variances, agreements or restrictions which would prevent, limit or otherwise restrict the construction or use of the Facility and that Owner will retain record legal title to such land until the construction of the Facility is completed. Owner has provided all necessary Site information with respect to soil conditions, existing buildings and structures, easements, restrictions and utility locations, and Contractor has, to the extent reasonable, relied upon such information as well as the aforementioned representations in designing the Facility and determining the Contract Price. Owner shall indemnify and hold harmless Contractor from any losses, expenses or damages Contractor may sustain in connection with the design or construction of the Facility because of differences between existing conditions and those conditions represented to Contractor.

B.   Owner has, prior to the date of this Contract, given to Contractor a copy of a site survey and report dated _____ and prepared by _____, relating to soil and subsurface conditions on or about the Site.

21.   Concealed and Unknown Conditions. Should Concealed and/or Unknown conditions be encountered in the performance of the Work below the surface of the ground or should Concealed and/or Unknown conditions in an existing structure be at variance with the conditions indicated by the TSD, the _____ report of _____ or other information furnished by Owner or should Concealed and/or Unknown physical conditions below the surface of the ground or Concealed and/or Unknown conditions in an existing structure of an unusual nature, differing materially from those ordinarily encountered and generally recognized as inherent in work of the character provided for in this Contract, be encountered, the Contract Price, Contractor's Fee and the Milestone Dates shall be equitably adjusted in accordance with paragraph 16 upon claim by Contractor within a reasonable time after the first observance of such conditions.

22.   Title to Work. Title to all Work completed or in the course of construction and paid for by Owner and title to all materials on account of which payment has been made by Owner shall vest in Owner upon payment, free and clear of any liens, encumbrances or rights of third parties.

23.   Liability of Contractor.

A.   Contractor shall indemnify, hold harmless and defend Owner and the agents and employees of Owner from and against all claims, damages, losses and expenses, including attorneys' fees, arising out of or resulting from the performance of the Work, provided that any such claim, damage, loss or expense (a) is attributable to bodily injury, sickness, disease or death or to injury to or destruction of tangible property (other than the Work itself) and (b) is caused in whole or in part by any negligent act or omission of Contractor, anyone directly employed by Contractor or anyone for whose acts Contractor may be liable, regardless of whether or not it is caused in part by a party indemnified hereunder. Further, Contractor shall indemnify and hold harmless Owner and the agents and employees of Owner from and against all claims, damages, losses and expenses, including attorneys' fees, arising out of or resulting from any negligent and/or wrongful acts, errors and/omissions committed in connection with the design of the Work by Contractor, anyone directly or indirectly employed by Contractor or anyone for whose acts Contractor may be liable. In the event that any attachments or liens shall be placed upon the property of Owner to secure or enforce any such claims, Contractor shall upon becoming aware of any such attachments or liens cause such attachments or liens to be dismissed or discharged.

B.   Contractor shall indemnify, hold harmless and defend Owner and the agents and employees of Owner from and against any claim by any employee of Contractor or any employee of any party retained by, through or under Contractor in connection with the Work if and to the extent that such claim arises out of or results from the performance of the Work and is not caused by any negligent and/or wrongful act or omission of Owner. In any and all claims against Owner or any agents or employees of Owner by any employee of Contractor or any employee of any party retained by, through or under Contractor in connection with the Work, the indemnification obligations of Contractor under this paragraph shall not be limited in any respect by any limitation on the amount or type of damages, compensation or benefits payable by or for Contractor or by or for any party retained by, through or under Contractor in connection with the Work under worker's compensation acts, disability benefit acts or other employee benefit acts.

C.   Contractor shall, subject to the receipt by Contractor of payment hereunder for any services rendered, work performed and/or materials furnished in connection with the design and/or construction of the Work, indemnify, hold harmless and defend Owner from and against any liens affecting the Facility and arising out of any such services rendered, work performed and/or materials furnished. If and to the extent that any liens affecting the Facility arise out of any such services rendered, work performed and/or materials furnished, Contractor shall, upon becoming aware of any such liens, cause such liens to be dismissed or discharged.

D.   Contractor shall promptly pay any and all sales, use and similar taxes related to the Work for which Contractor is liable imposed by any governmental authority and shall promptly pay any and all contributions, assessments and taxes for such items as unemployment compensation, social security and welfare and other benefits insofar as such contributions, assessments or taxes are based on compensation or other remuneration paid to employees of Contractor and are included in the Cost of the Work.

Contractor agrees to indemnify, hold harmless and defend Owner from and against all claims, damages, losses and expenses, including attorneys' fees, arising out of or resulting from the failure of Contractor either to pay such sales, use or similar taxes or to pay such contributions, assessments or taxes.

E.   The liability of Contractor to Owner under paragraphs 23A-D shall survive any termination of this Contract, and Contractor agrees that this paragraph shall remain in effect notwithstanding any such termination.

24.   <u>Waiver of Breach</u>. The failure of Owner or Contractor at any time to require performance of any provision of this Contract with respect to a particular matter shall in no way affect the right of Owner or Contractor to enforce such provision at a later date with respect to another matter, nor shall the waiver by Owner or Contractor of any breach or non-performance of any provision hereof be taken or held to be a waiver with respect to any succeeding breach or non-performance of such provision or as a waiver of the provision itself. Notwithstanding the foregoing, the making of final payment hereunder shall constitute a waiver of all claims by Owner except those arising from:

1. unsettled liens or claims;
2. faulty or defective work;
3. failure of the Work to comply with the requirements of this Contract;
4. terms of any special guarantees required by this Contract;
5. terms of the warranty set forth in paragraph 10;
6. latent defects; or
7. insurance and indemnification obligations of Contractor set forth in this Contract.

The acceptance of final payment shall constitute a waiver of all claims by Contractor except those previously made in writing and still unsettled.

25.   <u>Contractor's Liability Insurance</u>.

A.   Contractor shall maintain such insurance as will protect Owner and Contractor from those claims set forth in items 1 through 6 below which may arise out of or result from the operations of Contractor under this

Contract, whether such operations be by Contractor, by anyone directly or indirectly employed by Contractor or by anyone for whose acts Contractor may be liable:

1. claims under workmen's compensation, disability benefit and other similar employee benefit acts;

2. claims for damages because of bodily injury, occupational sickness, disease or death of any employees of Contractor;

3. claims for damages because of bodily injury, sickness, disease or death of any person other than an employee of Contractor;

4. claims for damages because of injury to or destruction of tangible property (other than the Work itself);

5. claims for damages arising after the construction of the Work is completed; or

6. claims for damages arising out of any negligent and/or wrongful acts, errors and/or omissions committed in connection with the design of the Work by Contractor.

B.   The insurance required to be maintained by Contractor shall not be written for less than any limits of liability specified in the TSD or less than any limits required by law, whichever is greater. Notwithstanding the foregoing, the worker's compensation insurance required to be maintained hereunder by Contractor shall be in an amount not less than any amount required by law, the employer's liability insurance required to be maintained hereunder by Contractor shall be in an amount not less than $5,000,000 on account of any one occurrence and the commercial general liability insurance, including coverage for premises operations (including "XCU" coverage), independent contractors, products and completed operations, contractual liability, personal injury and advertising liability, property damage (broad form) and automobile liability for all owned, non-owned and hired automobiles, required to be maintained by Contractor shall be in an amount not less than $5,000,000 on account of any one occurrence and not less than $5,000,000 on account of annual aggregate occurrences.

Contractor shall continue to maintain products and completed operations insurance coverage in at least the amounts set forth with respect to commercial general liability insurance coverage for the period beginning upon the commencement of the Work and ending one year after the construction of the Work has been substantially completed. The insurance required to be maintained by Contractor may be written under an umbrella form, but shall not be written with a deductible greater than $10,000 on account of any one occurrence or loss.

C.   Certificates evidencing that Contractor has obtained the insurance required to be maintained by Contractor shall be filed with Owner prior to commencement of construction of the Work. Such certificates shall indicate that Owner has been named as an additional insured with respect to all of such coverages, other than coverages relating to employer's liability and worker's compensation insurance, and shall contain a provision that such coverages shall not be materially changed or canceled until at least thirty (30) days' prior written notice has been given to Owner.

26.   Owner's Liability Insurance. Owner shall be responsible for purchasing and maintaining its own liability insurance and, at its option, may purchase and maintain such insurance as will protect it against claims, which may arise from operations under this Contract.

27.   Property Insurance.

A.   Owner shall purchase and maintain property insurance upon the entire Work at the Site to the full insurable value thereof, it being specifically understood that such property insurance shall not cover any tools or equipment at the site of the Work that belong to Contractor or any party retained by, through or under Contractor in connection with the Work and that are not wholly incorporated or required to be wholly incorporated into the Work. Such insurance shall be written with a deductible of Ten Thousand Dollars ($10,000) or less on account of any one occurrence or loss, shall include the interests of Owner, Contractor, subcontractors and sub-subcontractors in the Work, shall insure against all perils normally covered in an "all risk" form and shall contain an acknowledgment by the insurer which bars all rights of subrogation against the parties insured, it being specifically understood that all rights of

Contractor and Owner against each other for damages caused by fire or other perils to the extent covered by insurance provided under this paragraph 27 are waived.

B.   A certificate evidencing that Owner has obtained such property insurance and a copy of such property insurance policy shall be filed with Contractor prior to commencement of construction of the Work. Such certificate and policy shall be issued by a company or companies and be in form and substance reasonably satisfactory to Contractor and shall contain a provision that coverages afforded under such policy shall not be materially changed or canceled until at least thirty (30) days' prior written notice has been given to Contractor. If Owner does not intend to purchase such property insurance, Owner shall inform Contractor in writing prior to commencement of the Work, and Contractor may then effect such property insurance as will protect the interests of Contractor, subcontractors and sub-subcontractors in the Work, the cost of such property insurance in such case being an addition to the Contract Price. If Contractor is damaged by the failure of Owner to purchase or maintain such property insurance, then Owner shall bear all damages, loss and expenses properly attributable thereto.

28.   Accounting Records. Contractor shall check all materials, equipment and labor entering into the Work and shall keep such full and detailed accounts as may be necessary for proper financial management under this Contract in accordance with generally-accepted accounting standards and practices. Owner and/or Owner's representatives shall be afforded access at all reasonable times to all of the records, books, correspondence, instructions, drawings, receipts, vouchers, memoranda and similar data of Contractor relating to this Contract, and Contractor shall preserve all such records for a period of two years after Owner has made the final payment hereunder.

29.   Right of Owner to Terminate Contract.

A.   Owner may, without prejudice to any other right or remedy of Owner, and provided that Contractor has not eliminated such cause for terminating the Contract within fifteen (15) days after written notice thereof by Owner, terminate the Contract and take possession of the Work and of all drawings, specifications, permits, materials, tools and equipment at the Site and finish the Work by whatever method Owner may deem expedient:

1. if Contractor is adjudged bankrupt; or

2. if Contractor makes a general assignment for the benefit of creditors; or

3. if Contractor in a material respect refuses or fails, except in cases for which an extension of time is provided hereunder, to supply enough workmen or materials to reasonably assure Full Completion of the Work within the time provided by this Contract; or

4. if Contractor fails without reasonable cause to make payment either to subcontractors or for labor, materials or other work; or

5. if Contractor in a material respect disregards laws or ordinances or the instructions of Owner given in a valid manner pursuant to the Contract; or

6. if Contractor is otherwise guilty of a substantial violation of a provision of the Contract or breaches the Contract.

B.   If Owner terminates the Contract as provided above, Contractor shall only be entitled to receive payment if the sum of the expense incurred by Owner to finish the Work, including costs of additional design, engineering, managerial, administrative and construction services ("Owner's Costs") and the Cost of the Work and Contractor's Fee incurred by Contractor before termination ("Contractor's Costs") is less than the Contract Price. In such case, Owner shall promptly pay to Contractor Contractor's Costs (less any amount that Owner has previously paid to Contractor). However, if the sum of Owner's Costs and any and all payments made hereunder by Owner to Contractor for or in respect to the Contractor's Costs exceed the Contract Price, then Contractor shall promptly pay such excess to Owner upon demand. If requested by Contractor, the costs incurred by Owner as herein provided shall be certified by a certified public accountant. If requested by Owner, Contractor's Costs shall be certified by a certified public accountant.

C.   When Owner notifies Contractor of termination, Contractor will promptly remove any tools and equipment remaining on the Site, which Owner requests Contractor to remove, that are the property of Contractor, its subcontractors and suppliers. If Contractor fails to promptly respond, Owner has the right to remove the tools and equipment and bill Contractor for the cost of removal.

30.   <u>Right of Contractor to Stop Work or Terminate Contract</u>. Not withstanding anything to the contrary set forth in the Contract, if (i) the Work should be stopped under an order of any court or other governmental authority for a period of thirty (30) days through no act or omission of Contractor or of anyone employed by or subcontracted by Contractor or Contractor's representative or agent; or (ii) Owner should fail to pay Contractor any amount due and payable hereunder by Owner to Contractor and not disputed by Owner in good faith hereunder within fifteen (15) days after receipt of any application for progress payment and within thirty (30) days after receipt of any final application for payment submitted in accordance with this Contract; and (iii) the above-described causes for stopping the Work or terminating the Contract have not been eliminated by Owner within fifteen (15) days after written notice thereof by Contractor; then Contractor may stop the Work or terminate the Contract and recover from Owner payment for Contractor's Costs and for any proven reasonable loss actually incurred by Contractor as the result of cancellation fees or restocking fees. Further, in the event that Owner has failed to pay Contractor any amount due and payable by Owner to Contractor and not disputed by Owner in good faith within fifteen (15) days after receipt of any application for progress payment and within thirty (30) days after receipt of any final application for payment submitted in accordance with the Contract, then Contractor may assess a late charge against Owner equal to one and one-half percent (1 ½ %) of said payment for each month, or portion thereof, for which said payment is delinquent, said assessment being in addition to the Contract Price.

31.   <u>Dispute Costs</u>. If either Owner or Contractor commences any legal action against the other to enforce any of the terms or provisions of this

Contract, the prevailing party shall be entitled to recover from the other party all costs and expenses, including attorneys' fees, expert witness fees and expenses, fact witness fees and expenses, all copying costs, exhibit costs, mailing and delivery costs, long distance telephone costs, and lodging and travel costs incurred in relation to such legal action. The term "legal action" as used herein shall mean any lawsuit, arbitration or other legal proceeding before any court, tribunal or panel, including any appeals from a lower court decision, order or judgment. Any and all monetary damages, costs and fees awarded in any decision, order or judgment for any violation of the terms or provisions of this Contract shall bear interest from the date of such decision, order or judgment until paid at the rate of one and one-half percent (1 $\frac{1}{2}$%) per month until paid.

32.   Contract Modifications. No waiver, alteration or modification of any of the provisions of this Contract shall be binding upon either Owner or Contractor unless it shall be in writing and signed by both Owner and Contractor.

33.   Notices. All communications in writing between Owner and Contractor, including without limitation, applications for payment, shall be deemed to have been received by the addressee if delivered to the person for whom such communications are intended or if sent by certified mail, return receipt requested, or by telegram or by an express mail service addressed as follows:

If to Owner:

If to Contractor:

For the purpose of directions, the representative of Contractor shall be _____ and the representative of Owner shall be _____, unless otherwise specified in writing.

34.   Assignment. Neither Owner nor Contractor shall assign this Contract or sublet obligations hereunder in respect to the Work as a whole without the written consent of the other.

35.   Headings. The headings contained herein are inserted only as a matter of convenience and reference and are not meant to define, limit or describe the scope or intent of the TSD or in any way to affect the terms and provisions of this Contract.

36.   Applicable Law. The terms and provisions of this Contract and the rights and obligations of the parties shall be construed in accordance with the laws of the State of _____, without regard to any conflicts of laws.

37.   Succession of Rights and Obligations.   All rights and obligations under this Contract shall inure to and be binding upon the respective successors and assigns of Owner and Contractor.

IN WITNESS WHEREOF, Owner and Contractor have, by their duly authorized representatives, executed this Contract, in duplicate, as of the day, month and year first written above.

"OWNER"   "CONTRACTOR"

_____   _____

By: _____   By: _____

Printed Name: _____   Printed Name: _____

Title: _____   Title: _____

APPENDIX 1

Contract Term Definitions

A.   "Balance of Plant" means all equipment that Contractor will supply as specified in the TSD.

B.   "Balance of Plant Operational" means that Owner and Contractor reasonably agree that all Critical Systems are capable of safely providing services for commencement of commercial operation of the completely erected Units on a _____ (_____) hour a day basis. Critical Systems include step-up transformers, instrument air, natural gas, raw and demineralized water, interconnect wiring for Owner's Equipment (as herein after defined), the Programmable Logic Controller Central Processing Units and local inputs/outputs are operational for safe interlocks, and other necessary Unit support equipment.

C.   "Beyond the Reasonable Control of Contractor" as used in respect to any cause or circumstance means that such cause or circumstance could not have been reasonably foreseen or anticipated by Contractor on or before the date of the Contract or avoided by Contractor subsequently through the exercise of due diligence.

D.   "Concealed and/or Unknown" means that (i) the actual presence of such condition could not prior to the date of the Contract, be viewed by Contractor by means of a field inspection of the Site; (ii) Contractor, prior to the date of the Contract, had no notice or knowledge of and was not aware of the presence of such condition; and (iii) Contractor could not, in accordance with generally accepted engineering standards and practices of a business engaged in the power facility engineering, design, procurement and construction business and based upon information available to Contractor prior to the date of the Contract, have reasonably foreseen the presence of such condition.

E.   "Contractor's Fee" means an amount equal to fifteen and one-half percent (15 ½ %) of the total dollar value of the Cost of the Work as adjusted under the terms of this Contract.

F.   "Cost of the Work" means costs reasonably and necessarily incurred in the proper performance of the Work and paid or to be paid by Contractor. Such costs shall be at rates not higher than the standard paid in the locality of the Work, except with the consent of Owner, and shall include the items set forth below in items 1 through 16:

1.   compensation paid for labor, including but not limited to engineering, design and construction labor, in the direct employ of Contractor in the performance of the Work, including such insurance or other benefits, if any, as may be payable with respect thereto;

2.   compensation of employees of Contractor when stationed at the field office in whatever capacity employed, it being specifically understood and agreed that employees of Contractor, including the general manager and project manager of Contractor, engaged in managing, administering or accounting for the Work or in expediting the production or transportation of materials or equipment incorporated into the Work shall be considered as stationed at the field office and their compensation paid for that portion of their time spent on the Work;

3.   cost of contributions, assessments or taxes for such items as unemployment compensation and social security insofar as such cost is based on compensation or other remuneration paid to employees of Contractor included in the Cost of the Work;

4.   reasonable transportation, traveling, meal and hotel expenses of Contractor's employees incurred in discharging duties connected with the Work;

5.   cost of all materials, supplies and equipment incorporated into the Work, including reasonable costs of transportation thereof;

6.   expenses incurred pursuant to this Contract by Contractor in respect to subcontractors, materialmen and/or other parties retained by, through or under Contractor for Work required to be performed to design or construct the Facility;

7.   cost, including transportation and maintenance, of all materials, including all consumables and expendables used in the engineering and design of the Facility, and all supplies, equipment, temporary facilities and hand tools which are not owned by workmen and which are consumed in the performance of the Work, and cost less salvage value in respect to such items which are used but not consumed and which remain the property of the Contractor;

8.   rental charges of all necessary machinery and equipment, exclusive of hand tools, used at the Site, whether rented from Contractor or others, including installation, minor repairs and replacements, dismantling, removal, transportation and delivery costs thereof, at rental charges consistent with those prevailing in the area of the Work;

9.   cost of premiums for all bonds, guarantees, warranties and insurance which Contractor is required to purchase and/or maintain in order to carry out the Work;

10.   sales, use or similar taxes related to the Work and for which Contractor is liable imposed by any governmental authority;

11.   permit fees, royalties, damages for infringement of patents and costs of defending suits therefor and deposits lost for causes other than the negligence of Contractor;

12.   expenses of reconstruction and/or costs to replace and/or repair damaged materials and supplies provided that Contractor is not compensated for such expenses and/or costs by insurance otherwise and that such expenses and/or costs have resulted from causes other than the fault or omission of Contractor, it being specifically understood and agreed that such expenses and/or costs shall include costs incurred as a result of any act or neglect of Owner or of any employee, agent or representative of Owner, any act or neglect of any separate contractor employed by Owner and/or any casualty or so-called "war-risk";

13. expenses for such items as shipping, delivery, facsimile transmissions, telegrams, long distance communications, telephone service at the Site, postage, expressage, copying, computer usage and petty cash items purchased in connection with the Work;

14. cost of removal of all debris from the Site;

15. cost incurred due to an emergency affecting the safety of persons and property; and

16. other costs incurred in the performance of the Work if and to the extent such costs are approved in writing by Owner.

Except as otherwise provided in this Contract, Cost of Work shall not include any of the items set forth below in items 1 through 4:

1. salaries or other compensation of the marketing executives, general managers, or auditors of Contractor at the principal office or branch offices of Contractor, unless assigned temporarily to the project;

2. expenses of the principal or branch offices of Contractor other than the field office at the Site;

3. any part of the capital expenses of Contractor, including interest on capital of Contractor employed for the Work; and

4. general administrative or other indirect overhead expenses of any kind, except as may be expressly included above in items 1 through 16.

G. "Facility" means the type of facility as described and defined in the TSD.

H. "Full Completion," "Fully Complete" and "Fully Completed" mean that all of the Work, including punch list items, has been completed in accordance with the Contract.

I.   "Minimal Interference" means all labor and materials necessary to fully complete the Work, and access to the entire Facility by Contractor and any party retained by, through or under Contractor in connection with the Work, subject to Owner's operation schedule but not to be less than eight (8) hours per week day, and sixteen (16) hours per day on Saturdays and Sundays.

J.   "Punch List" means a written list to be jointly prepared by Owner and Contractor when the Work, or a designated portion of the Work, is Substantially Completed that shall include items to be completed or corrected by Contractor.

K.   "Site" means the site to be provided by the Owner as described and defined in the TSD.

L.   "Substantial Completion," "Substantially Complete" and "Substantially Completed" mean that all the Work has been completed in accordance with the Contract, including the acceptance of turnover documents by Owner, sufficiently to enable Owner, subject to and upon the performance of any separate work to be done by Owner and/or by any party, other than Contractor, retained by, through or under Owner, to take possession of the applicable Unit and to commence commercial operations with that Unit with Minimal Interference to that Unit.

M.   "TSD" means the Technical Scope Document, which is attached and incorporated herein as Exhibit A.

N.   "Turnover Documents" mean manuals prepared by Contractor for each plant system (as described in the TSD) detailing the testing procedures to confirm readiness for commercial operation and turnover of control of the Facility systems (as described in the TSD) to Owner.

O.   "Units," and individually "Unit," mean the _____.

P.   "Weather Day" or "Weather Days" mean any day or part of a day on which performance of any part of the Work necessary to complete the Facility is prohibited or significantly impaired due to precipitation, lightning or wind or the effects of precipitation, lightening or wind at the location of the Work.

Q.   "Work" means all labor, materials, supplies, equipment and services necessary to procure material, engineer, design and construct the Facility as shown and described in the TSD and perform all other services as specified in the TSD.

EXHIBIT A

TECHNICAL SCOPE DOCUMENT

EXHIBIT B

PARTIAL WAIVER OF LIENS
AND INDEMNITY AGREEMENT

KNOW ALL MEN BY THESE PRESENTS THAT _____ (referred to as "Contractor") has rendered services to, performed work for and/or furnished materials to _____ (referred to as "Owner") in connection with the design and/or construction of a _____ facility situated at _____.

NOW, THEREFORE, for one dollar ($1.00) and other good and valuable consideration, the receipt and sufficiency of which is hereby acknowledged, Contractor does hereby, subject to the reservation set forth below and upon receipt of _____ Dollars ($_____) representing the amount due and payable by Owner to Contractor for services rendered, work performed and/or materials furnished by Contractor in connection with the design and/or construction of the facility during the period commencing _____ and ending _____, waive and release all liens or rights of lien which Contractor has pursuant to the laws of the State of _____ by virtue of services rendered, work performed and/or materials furnished during such period in connection with the design and/or construction of the facility.

Contractor does hereby reserve, and does not hereby waive or release, any lien or right of lien which Contractor has pursuant to the laws of the State of _____ for services rendered, work performed and/or materials furnished by Contractor after the stated period in connection with the design and/or construction of the facility.

Contractor does hereby, subject to the receipt of the amount due and payable by Owner to Contractor for services rendered, work performed and/or materials furnished by Contractor in connection with the design and/or construction of the facility during the stated period agree to indemnify, hold harmless and defend Owner from and against any liens and any associated claims, disputes, lawsuits, and expenses affecting the facility and filed by any party retained by, through or under Contractor in respect to services rendered, work performed and/or materials furnished during the stated period in connection with the design and/or construction of the facility.

This Partial Waiver of Liens and Indemnity Agreement shall be binding upon Contractor and the successors and assigns of Contractor and shall inure to the benefit of Owner and the successors and assigns of Owner.

IN WITNESS WHEREOF Contractor has executed this Partial Waiver of Liens and Indemnity Agreement this _____ day of _____, 20____.

_____

By: _____

Its: _____

STATE OF _____

COUNTY OF _____

This instrument was acknowledged before me on_____, 20____ by _____ as _____ of _____.

[S E A L]

_____

Notary Public

My commission expires: _____

EXHIBIT C

FINAL WAIVER OF LIENS AND INDEMNITY AGREEMENT

KNOW ALL MEN BY THESE PRESENTS THAT _____ (referred to as "Contractor") has rendered services to, performed work for and/or furnished materials to _____ (referred to as "Owner") in connection with the design and/or construction of a _____ facility situated at _____.

NOW, THEREFORE, for one dollar ($1.00) and other good and valuable consideration, the receipt and sufficiency of which is hereby acknowledged, Contractor does hereby, upon receipt of _____ Dollars ($_____) representing the final payment due and payable by Owner to Contractor for services rendered, work performed and/or materials furnished by Contractor in connection with the design and/or construction of the facility, waive and release all liens or rights of lien which Contractor has pursuant to the laws of the State of _____ by virtue of services rendered, work performed and/or materials furnished in connection with the design and/or construction of the facility.

Contractor does hereby, subject to the receipt of the aforementioned final payment due and payable by Owner to Contractor for services rendered, work performed and/or materials furnished by Contractor in connection with the design and/or construction of the facility, agree to indemnify, hold harmless and defend Owner from and against any liens affecting the aforesaid facility and filed by any party retained by, through or under Contractor in respect to services rendered, work performed and/or materials furnished in connection with the design and/or construction of the facility.

This Final Waiver of Liens and Indemnity Agreement shall be binding upon Contractor and the successors and assigns of Contractor and shall inure to the benefit of Owner and the successors and assigns of Owner.

IN WITNESS WHEREOF Contractor has executed this Final Waiver of Liens and Indemnity Agreement this _____ day of _____, 20___.

_____

By: _____

Its: _____

STATE OF _____

COUNTY OF _____

This instrument was acknowledged before me on_____, 20___ by _____ as _____ of _____.

[S E A L]

_____

Notary Public

My commission expires: _____

Courtesy of Trent A. Gudgel, Hall, Estill, Hardwick, Gable, Golden & Nelson PC

APPENDIX C

MECHANICS' AND MATERIALMEN'S LIEN

STATE OF OKLAHOMA
COUNTY OF _____

KNOW ALL MEN BY THESE PRESENTS: that [subcontractor lien claimant], a limited liability company, with its principal place of business at [address], has a claim against [general contractor] for the sum of $_____, plus interest as provided by Oklahoma law, due to it and that the claim is made for and on account of labor performed and material furnished by it between [dates during which work that is subject of lien was performed], for the improvement of certain real property more particularly described below, according to the itemized statement attached hereto and incorporated herein as Exhibit A. [Subcontractor lien claimant]caused a pre-lien notice to be served on the general contractor, [general contractor name and address], and the property owner, [property owner name and address], as required by Oklahoma law and as verified by the affidavit attached hereto and incorporated herein as Exhibit B. Labor was performed and material was furnished upon the buildings, fixtures, appurtenances and premises known as [project name] located at [project address] and more specifically described as:

[project legal description]

The sum is just, due and unpaid, and [subcontractor lien claimant] has a claim and lien upon the improvements, buildings and appurtenances and upon the premises on which the same are situated in the amount of $_____, plus interest, as above set forth, according to the laws of the State of Oklahoma.

Dated this _____ day of _____, 20___.

[name of lien claimant]

By: _____

SUBSCRIBED AND AFFIRMED before me, a Notary Public, on this
_____ day of _____, 20__.

[S E A L]

Notary Public: _____

My commission expires: _____

My commission number: _____

EXHIBIT A

ITEMIZED STATEMENT

EXHIBIT B

PRE-LIEN NOTICE AFFIDAVIT

STATE OF OKLAHOMA
COUNTY OF _____

The undersigned, [name of affiant], of lawful age, being first duly sworn upon oath, deposes and states as follows:

1.   I am [job title] for [subcontractor lien claimant], a limited   liability company, whose principal place of business is located at [address].

2.   On _____, 20___, I caused pre-lien notices to be sent, via certified mail, return receipt requested to [general contractor] and [owner] notifying them of the material, equipment and labor furnished by [subcontractor lien claimant] for [describe project] at [project address], and more specifically described herein above.

FURTHER AFFIANT SAYETH NOT.

Dated this _____ day of _____, 20___.

SUBSCRIBED AND AFFIRMED before me, a Notary Public, on this _____ day of _____, 20__.

_____

[S E A L]

Notary Public: _____

My commission expires: _____

My commission number: _____

MAIL NOTICE OF MECHANIC'S LIEN TO:

General Contractor: [General Contractor Name and Address]

Property Owner: [Owner name and Address]

Courtesy of Trent A. Gudgel, Hall, Estill, Hardwick, Gable, Golden & Nelson PC

APPENDIX D

PRE-LIEN NOTICE FORM

VIA CERTIFIED MAIL NO.   [DATE]
[Property Owner's name]
[Address]

VIA CERTIFIED MAIL NO.
[General Contractor's name]
[Address]

Re: Pre-Lien Notice

Project: [project description]

TO WHOM IT MAY CONCERN:

This letter serves as a pre -lien notice. [Lien claimant name and address] has furnished labor, materials and equipment between [date work started] and [date work ended], in excess of $2,500.00 [Oklahoma statutory requirement] to [general contractor name and address], relating to the construction of [generally describe lien claimant's work] for [identify project and project address]. The total amount owed to [lien claimant] has not been paid.

Under Oklahoma law, [general contractor] is required to provide any lien claimant with the name and last known address of the owner of the subject property within five days upon request. See 42 O.S. 2001, ' 142.6(B)(6). In the event the name and address of the owner of the subject property differs from that set forth above, [general contractor] is hereby requested to notify the undersigned of the correct name and address of the owner of record, in writing, no later than five days after the date of this pre-lien notice.

Sincerely yours,

_____

Courtesy of Trent A. Gudgel, Hall, Estill, Hardwick, Gable, Golden & Nelson PC

APPENDIX E

NORTH CAROLINA LAW ON
INDEMNITY AND CONTRIBUTION

Two legal issues that arise most often in construction litigation are the concepts of indemnity and contribution. These concepts are often the key issues that must be resolved either in mediation or at trial in order to allocate liability between multiple defendants. This paper is intended to act as a brief introduction to these issues, and to serve as a reference for industry professional and insurance adjusters when they are confronted with a claim involving these issues.

INDEMNITY

The purpose of indemnity is to make good and save another harmless from loss on some obligation which he has incurred or is about to incur to a third party. *New Amsterdam Casualty Co. v. Waller*, 233 N.C. 536, 537, 64 S.E.2d 826, 827-28 (1951). In North Carolina, a party's right to indemnification can rest on three bases: (1) an express contract; (2) a contract implied-in-fact; or (3) equitable concepts arising from the tort theory of indemnity, often referred to as a contract implied-in-law. *Kaleel Builders, Inc. v. Ashby*, 161 N.C. App. 34, 38, 587 S.E.2d 470, 474 (2003).

A.   Contractual Indemnity

The most common form of indemnity that arises in construction matter is an express indemnity provision contained in a construction contract. There are numerous examples of such provisions, and indemnity agreements are often contained in commonly used form contracts. For example, the following provision is found in American Institute of Architects ("AIA") Document A201, General Conditions of the Contract for Construction at Section 3.18.1:

> To the fullest extend permitted by law and to the extent claims, damages, losses or expenses are not covered by Project Management Protective Liability insurance purchased by the Contractor in accordance with Paragraph

11.2, the Contractor shall indemnify and hold harmless the Owner, Architect, Architect's consultants, and agents and employees of any of them from and against claims, damages, losses and expenses, including but not limited to attorney's fees, arising out of or resulting from performance of the Work, provided that such claim, damage, loss or expense is attributable to bodily injury, sickness, disease or death, or to injury to or destruction of tangible property (other than the Work itself), but only to the extent caused by the negligent acts or omissions of the Contractor, a Subcontractor, anyone directly or indirectly employed by them or anyone for whose acts they may be liable, regardless of whether or not such claim, damage, loss or expense is caused in part by a party indemnified hereunder. Such obligation shall not be construed to negate, abridge, or reduce other rights or obligations of indemnity which would otherwise exist as to a party or person described in this Paragraph 3.18.

Similar provisions are contained in Section 4.6.1 of AIA Document A401, the Contractor-Subcontractors Agreement:

To the fullest extent permitted by law, the Contractor shall indemnify and hold harmless the Subcontractor, the Subcontractor's Sub-subcontractors, and agents and employees of any of them from and against claims, damages, losses and expenses, including but not limited to attorney's fees, arising out of or resulting from performance of the Work in the affected area if in fact the material or substance presents the risk of bodily injury, sickness, disease or death, or to injury to or destruction of tangible property (other than the Work itself) including loss of use resulting therefrom and provided that such damage, loss or expense is not due to the sole negligence of a party seeking indemnity.

To the fullest extent permitted by law, the Subcontractor shall indemnify and hold harmless the Owner, Contractor,

Architect, Architect's consultants, and agents and employees of any of them from and against claims, damages, losses and expenses, including but not limited to attorney's fees, arising out of or resulting from performance of the Subcontractor's Work under this Subcontract, provided that any such claim, damage, loss or expense is attributable to bodily injury, sickness, disease or death, or to injury to or destruction of tangible property (other than the Work itself), but only to the extent caused by the negligent acts or omissions of the Subcontractor, the Subcontractor's Sub-subcontractors, anyone directly or indirectly employed by them or anyone for whose acts they may be liable, regardless of whether or not such claim, damage, loss or expense is caused in party by a party indemnified hereunder. Such obligation shall not be construed to negate, abridge, or otherwise reduce other rights or obligations of indemnity which would otherwise exist as to a party or person described in this Paragraph 4.6.

Another example of indemnity is contained in the Model General Agreement of Indemnity often used between a surety and a principal or indemnitor:

1. Indemnification

The Undersigned shall indemnify and keep indemnified the Surety against any and all liability for losses and expenses, of whatsoever kind or nature, including the fees and disbursements of counsel, and against any and all said losses and expenses, which the Surety may sustain or incur, (i) by reason of having executed or procured the execution of any Bond or Bonds, (ii) by reason of the failure of the Undersigned to perform or comply with the covenants and conditions of this Agreement, or (iii) in enforcing any of the covenants and conditions of this Agreement.

All of these provisions typically would be enforced under North Carolina law.

In an indemnity contract, the agreement will be construed to cover all losses, damages, and liabilities that reasonably appear to have been within the contemplation of the parties, but not those which are neither expressed nor reasonably inferable from the terms. *Kirkpatrick & Associates, Inc. v. Wickes Corp.* 53 N.C. App. 306, 308, 280 S.E.2d 632, 634 (1981). Indemnity contracts are entered into to save one party harmless from some loss or obligation which it has incurred or may incur to a third party. *Id.*

North Carolina, through North Carolina General Statute § 22B-1, limits indemnification in construction-related contracts to a party's own negligence. In particular, this statute reads as follows:

> Any promise or agreement in, or in connection with, a contract or agreement relative to the design, planning, construction, alteration, repair or maintenance of a building, structure, highway, road, appurtenance or appliance, including moving, demolition or excavating connected therewith, purporting to indemnify or hold harmless the promisee, the promisee's independent contractors, agents, employees, or indemnitees against liability for damages arising out of bodily injury to persons or damage to property proximately caused by or resulting from the negligence, in whole or in part, of the promisee, its independent contractors, agents, employees, or indemnitees, is against public policy and is void and unenforceable. Nothing contained in this section shall prevent or prohibit a contract, promise or agreement whereby a promisor shall indemnify or hold harmless any promisee or the promisee's independent contractors, agents, employees or indemnitees against liability for damages resulting from the sole negligence of the promisor, its agents or employees. This section shall not affect an insurance contract, worker's compensation, or any other agreement issued by an insurer, nor shall this section apply to promises or agreements under which a

public utility as defined in G.S. 62-3(23) including a railroad corporation as an indemnitee. This section shall not apply to contracts entered into by the Department of Transportation pursuant to G.S. 136-28.1.

Parties may not require another party to a construction contract to indemnify them for their own negligence, as such provisions violate public policy in North Carolina. *International Paper Co. v. Corporate Constructors, Inc.*, 96 N.C. App 312, 385 S.E.2d 553 (1989). This statute does not, however, prevent parties from requiring their contractors or subcontractors to purchase insurance on their behalf or have them named as an additional insured party. Courts will not strike down an entire contract if it includes an over-broad indemnification clause; rather, the court will reform the contract to eliminate the illegal provisions. It should be noted that an overly broad indemnity provision in a construction contract will not be reformed by a court; rather, the entire indemnity provision will be removed from the contract. *Jackson v. Associated Scaffolders & Equipment Co.*, 152 N.C. App. 687, 568 S.E.2d 666 (2002).

If such a provision is contained in a contract not based in construction, the indemnification provision would be enforced, even if the indemnitor was required to indemnify the indemnitee for the indemnitee's own negligence.

B.   Implied Indemnity

A right of indemnity implied-in-fact stems from the existence of a binding contract between two parties that necessarily implies the right. The implication is derived from the relationship between the parties, circumstances of the parties' conduct, and that the creation of the indemnitor/indemnitee relationship is derivative of the contracting parties' intended agreement. *McDonald v. Scarboro*, 91 N.C. App. 13, 370 S.E.2d 680, *disc. review denied*, 323 N.C. 476, 373 S.E.2d 864 (1988). The situations where the relationship between the parties implies that indemnification exists are in surety and agency. In *Kaleel* discussed above, the Court of Appeals declined to state that implied-in-fact indemnification existed due solely to a contractor-subcontractor relationship, while at the same time stating

283

implied-in-fact indemnity could exist in a relationship other than surety or agency.

C.   Common Law Indemnity

While contractual indemnity implied-in-law is a rather discrete legal fiction, North Carolina appellate courts have been consistent in recognizing "implied-in-law" or common law indemnity. Common law indemnity arises from an underlying tort, where a passive tort-feasor pays the judgment owed by an active tort-feasor to the injured third party. The Supreme Court set this out clearly: "The old-time judges said that the duty imposed by law upon the actively negligent tort-feasor to reimburse the passively negligent tort-feasor for the damages paid by him to the victim of their joint tort was based on an implied contract, meaning a contract implied in law from the circumstance that the passively negligent tort-feasor had discharged an obligation for which the actively negligent tort-feasor was primarily liable. And this is all the courts mean today when they declare that the right of the passively negligent tort-feasor to indemnity from the actively negligent tort-feasor rests upon an implied contract. There is, of course, in such case no contract implied in fact. This is necessarily so because contracts implied in fact are true contracts based on consent." *Hunsucker v. High Point Bending & Chair Co.*, 237 N.C. 559, 563-64, 75 S.E.2d 768, 771 (1953).

In order for common law indemnity to apply, the defendants (or defendant and third party defendants) must be (1) jointly and severally liable to the plaintiff, and (2) either (a) one has been passively negligent but is exposed to liability through the active negligence of the other or (b) one alone has done the act that produced the injury but the other is derivatively liable for the negligence of the wrong doer. *Kaleel*, 161 N.C. App. at 41, 587 S.E.2d at 475 (quoting *Edwards v. Hammill*, 262 N.C. 528, 531 (1982)). While this principle often applies in personal injury actions, it typically does not apply in construction claims. For example, when both a general contractor and a subcontractor are sued, there are usually independent claims against each entity, and thus common law indemnity will not apply. *E.g., Sullivan v. Smith*, 56 N.C. App. 525, 531-32, 289 S.E.2d 870, 874 (1982)(contractor's duty to supervise precludes common law indemnity); *Hendricks v. Fay, Inc.*, 273 N.C. 59, 62, 159 S.E.2d 362, 365 (1968)(no derivative liability for negligence of an independent contractor).

As our Court of Appeals has noted, although there has been some confusion to the contrary, the law with respect to exculpatory clauses is different from that with respect to indemnification clauses. *Candid Camera Video World, Inc. v. Mathews*, 76 N.C. App. 634, 636, 334 S.E.2d 94, 95 (1985), *disc. review denied*, 315 N.C. 390, 338 S.E.2d 879 (1986). Specifically, there is a distinction between contracts whereby one seeks to wholly exempt himself from liability for the consequences of his negligent acts, and contracts of indemnity against liability imposed for the consequences of his negligent acts. *Gibbs v. Carolina Power & Light Co.*, 265 N.C. 459, 467, 144 S.E.2d 393, 400 (1965).

CONTRIBUTION

Contribution arises when more than one tort-feasor is liable for a single injury; it permits one tort-feasor to demand assistance from the other tort-feasor if his payment to the injured party exceeds his pro rata share of the damage. The right to contribution, where one person discharges more than his just share of a common burden, does not arise from contract, but has its foundation in, and is controlled by, principles of equity and natural justice. *Moore v. Moore*, 4 Hawks 358, 1826 WL 241, 1 (1826). As the North Carolina Supreme Court has put it, "equality is equity among those standing in same situation." *Harvey v. Oettinger* 194 N.C. 483, 140 S.E. 86 (1927).

The right to contribution is specifically set out by statute in North Carolina. North Carolina General Statutes §1B-1 reads as follows:

(a) Except as otherwise provided in this Article, where two or more persons become jointly or severally liable in tort for the same injury to person or property or for the same wrongful death, there is a right of contribution among them even though judgment has not been recovered against all or any of them.

(b) The right of contribution exists only in favor of a tort-feasor who has paid more than his pro rata share of the common liability, and his total recovery is limited to the amount paid by him in excess of his pro rata share. No tort-feasor is compelled to make contribution beyond his own pro rata share of the entire liability.

(c)   There is no right of contribution in favor of any tort-feasor who has intentionally caused or contributed to the injury or wrongful death.

(d)   A tort-feasor who enters into a settlement with a claimant is not entitled to recover contribution from another tort-feasor whose liability for the injury or wrongful death has not been extinguished nor in respect to any amount paid in a settlement which is in excess of what was reasonable.

(e)   A liability insurer, who by payment has discharged in full or in part the liability of a tort-feasor and has thereby discharged in full its obligation as insurer, succeeds to the tort-feasor's right of contribution to the extent of the amount it has paid in excess of the tort-feasor's pro rata share of the common liability. This provision does not limit or impair any right of subrogation arising from any other relationship.

(f)   This Article does not impair any right of indemnity under existing law. Where one tort-feasor is entitled to indemnity from another, the right of the indemnity obligee is for indemnity and not contribution, and the indemnity obligor is not entitled to contribution from the obligee for any portion of his indemnity obligation.

The right to contribution does not exist where the relationship between the parties arises out of contract, and thus contribution does not exist where the claims are based on breach of contract or breach of warranty. *Holland v. Edgerton*, 85 N.C. App. 567, 571, 355 S.E.2d 514, 517 (1987); *Land v. Tall House Building Co.*, 165 N.C. 880, 602 S.E.2d 1 (2004). Thus, in most construction claims, the remedy between the defendants will be for breach of contract or indemnification, rather than joint tortfeasor contribution. It is important to note that the rights of contribution and indemnity are mutually inconsistent; the former assumes joint fault, the latter only derivative fault. *Edwards v. Hamill*, 262 N.C. 528, 531, 138 S.E.2d 151, 153 (1964).

Another concept of indemnification that is problematic in cases where insurance coverage exists comes from the provision of General Statutes § 1B-1(b) when a plaintiff chooses to collect a judgment against one defendant only, as it may do under General Statutes § 1B-3. Because North Carolina is a contributory negligence state rather than a comparative negligence state, this can lead to impractical results as recognition of relative degree of fault between joint defendants is expressly prohibited by statute. N.C. Gen. Stat. § 1B-2. By way of example, assume there is a case where a large and solvent corporation and an insolvent individual are sued as joint tortfeasors. If a judgment is obtained against both of them, the plaintiff may choose which defendant to go after the judgment from. This means that the defendant could get the entire judgment from the solvent corporation, and the only recourse the corporation has is to seek contribution from the individual defendant for its pro rata share. In such a situation, the corporate defendant could end up paying the entire judgment, even though it may be only responsible for a very small percentage of fault.

Courtesy of William W. Pollock, Cranfill, Sumner & Hartzog LLP

APPENDIX F

SETTLEMENT AGREEMENT AND RELEASE OF ALL CLAIMS

STATE OF NORTH CAROLINA
COUNTY OF _____

THIS RELEASE OF ALL CLAIMS, given the _____ day of October, 2006, by and between _____, as Releasor, and _____, as Releasee.

WITNESSES

WHEREAS, Releasor contends that it has been suffered damages as a result of certain actions alleged to have been taken by Releasee in connection with the parties' work on the _____ project (the "Project") for _____ (the "Owner") in which Releasor acted as general contractor and Releasee acted as the engineer for the Owner; and

WHEREAS, Releasor instituted in the General Court of Justice, Superior Court Division of _____ County, North Carolina, a Civil Action seeking money damages for these damages, said Civil Action being entitled "_____" bearing Court File No. 05-CVS-___ (the "Civil Action"); and

WHEREAS, Releasee denies the allegations and contentions of Releasor as set forth in the Civil Action; and

WHEREAS, Releasor desires to resolve, settle, and compromise its claims against Releasee as hereinafter provided and subject to the terms hereinafter set forth; and

THEREFORE, in consideration of the foregoing premises and the execution of this document, and in consideration of payment of the sum of $_____ on behalf of Releasee, the receipt and sufficiency of which consideration is hereby expressly acknowledged, Releasor does contract, covenant and agree as follows:

1.   Warranties and Indemnification.

a.   Releasor warrants and represents to the best of its knowledge and belief that no other person or entity has any right, title or interest (i) in the claims arising out of the Civil Action; and (ii) in the monetary consideration paid hereunder.

b.   Releasor warrants and represents that it has been fully informed and has full knowledge of the terms, conditions and effects of this Settlement Agreement and Release of All Claims.

c.   Releasor warrants and represents that it has either personally or through an attorney fully investigated to its full satisfaction, all facts surrounding various claims, controversies and disputes and is fully satisfied with the terms and effects of this Settlement Agreement and Release of All Claims.

d.   Releasor warrants and represents that no promise or inducement has been offered or made except as herein set forth, nor has there been any representation made by any party or party's attorneys as to any tax effects or consequences resulting from the payment arising out or under this Settlement Agreement and Release of All Claims and that this Settlement Agreement and Release of All Claims is executed without reliance on any such statement or representation by any other party or party's attorney. Releasor further acknowledges that it shall be responsible for any and all tax consequences, if any, which may result from any payment under this Settlement Agreement and Release of All Claims.

2.   Full Satisfaction and Release.

Releasor stipulates and agrees that this compromise settlement is in full and complete satisfaction of all of its claims (real, potential, or otherwise) arising out of the Civil Action and any resulting injury or damage claimed on account of the incident set forth in the Civil Action. For and in consideration of the monetary consideration set forth above, Releasor, for itself and its heirs, administrators, executors, successors and assigns, does hereby fully and forever release, acquit and discharge _____
and its agents, servants, officers, directors, shareholders, members, partners,

employees, insurers, attorneys, subsidiary and related corporations, partnerships and entities wherever same may be incorporated or domesticated and any and all other persons and entities of and from any and all actions, causes of actions, damages, costs, liabilities, claims and/or demands of whatsoever kind or nature, known and unknown, suspected and unsuspected (specifically including but not limited to all claims set forth in the Civil Action and including consequential damages and any and all claims for witness fees, attorney's fees, or costs incurred in connection with the Civil Action) based upon, arising out of, or in anyway connected with the Project or the matters set forth in the Civil Action. Releasor hereby reserves any claims it may have against the _____in the Civil Action, and this Release shall in no way be construed as affecting such claims or as a release of any claim against the _____.

3.   Dismissal of the Civil Action.

Releasor covenants and agrees that upon receipt by its attorneys of the payment recited herein as settlement hereunder, it will execute contemporaneously with this Settlement Agreement and Release of All Claims a Notice of Voluntary Dismissal with Prejudice of the claims in the Civil Action against Releasee, an unexecuted copy of which is attached hereto as "Exhibit A," and shall file the same with the Clerk of Court of _____ County, North Carolina. Releasor further covenants and agrees that it will direct its attorneys to mail by first class mail to counsel for Releasee on the same day as execution of this Settlement Agreement and Release of All Claims the original, fully executed Settlement Agreement and Release of All Claims and a file-stamped copy of the fully executed Notice of Dismissal with Prejudice showing the filing of the same with the Clerk of Court.

4.   Indemnification.

Releasor agrees to indemnify and save harmless Releasee from and against all claims and demands whatsoever in account of or in any way growing out of the matters referred to in the Civil Action.

5.   Confidentiality of Settlement Agreement and Release of All Claims.

Releasor and Releasee agree that neither party nor their attorneys or representatives shall reveal to anyone, other than as may be mutually agreed to in writing or as required by any order of a court of competent jurisdiction, any of the terms of this Settlement Agreement and Release of All Claims or any of the amounts, numbers or terms and conditions of any sums payable hereunder.

6.   Miscellaneous Provisions.

a.   Additional Instruments. Releasor agrees that it shall execute and deliver such other and further notices, releases, acquittances, and other documents as may be necessary to implement fully the terms and provisions hereof.

b.   Agreement to Survive. All warrants, covenants, agreements and releases contained herein shall survive this Settlement Agreement and Release of All Claims.

c.   Governing Law. The validity, construction, interpretation, and administration of this Settlement Agreement and Release of All Claims shall be governed by the laws of the State of North Carolina.

d.   Entire Agreement. This Settlement Agreement and Release of All Claims constitutes the entire agreement between the parties pertaining to the subject matter contained herein, and its terms are contractual and not a mere recital.

e.   No Liability Admitted. Releasor understands and agrees that this is a compromise settlement of disputed claims and that acceptance of any benefits pursuant to this Settlement Agreement and Release of All Claims or the offer of such benefits is intended merely to terminate any and all claims by it and that acceptance of benefits under this Settlement Agreement and Release of All Claims or payments of such benefits is not to be construed as an admission of liability or an acquiescence by any party to the claims and contentions of any other party.

f.   <u>Invalid Provisions</u>. If, after the day hereof, any provision of this Settlement Agreement and Release of All Claims is held to be illegal, invalid, or unenforceable under present or future laws in effect during the term of this Settlement Agreement and Release of All Claims, such provisions shall be fully severable.

g.   <u>Execution</u>. Releasor represents and states that it has fully and carefully read this Settlement Agreement and Release of All Claims, that it knows the contends thereof, that all necessary corporate approvals for entering into this Agreement have been obtained, that the representative signing on behalf of Releasor has been duly authorized to execute this document on behalf of Releasor and that he or she has signed the same of his or her own free and voluntary act.

_____

BY: _____ (SEAL)

_____

Name

_____

Title

ATTEST:

_____

Secretary

STATE OF NORTH CAROLINA
COUNTY OF _____

I, _____, a Notary Public for said County and State, do hereby certify that _____ personally appeared before me this day and acknowledged the due execution of the foregoing instrument.

Witness my hand and official seal, this the _____ day of October 2006.

_____
NOTARY PUBLIC

My Commission Expires: _____

Courtesy of William W. Pollock, Cranfill, Sumner & Hartzog LLP

APPENDIX G

The General Conditions of the Contract for Construction, American Institute of Architects Document

The General Conditions of the Contract for Construction, American Institute of Architects Document A201 (1997), contain the following provisions:

4.3.1 **Definition.** A Claim is a demand or assertion by one of the parties seeking, as a matter of right, adjustment or interpretation of Contract terms, payment of money, extension of time or other relief with respect to the terms of the Contract. The term "Claim" also includes other disputes and matters in question between the Owner and Contractor arising out of or relating to the Contract. Claims must be made by written notice. The responsibility to substantiate Claims shall rest with the party making the Claim.

4.3.2 **Time Limits on Claims.** Claims by either party must be initiated within 21 days after occurrence of the event giving rise to such Claim or within 21 days after the claimant first recognizes the condition giving rise to the Claim, whichever is later. Claims must be made by written notice.

4.3.5 Claims for Additional Costs. If the Contractor wishes to make Claim for an increase in the Contract Sum, written notice as provided herein shall be given before proceeding to execute the Work. Prior notice is not required for Claims relating to an emergency endangering life or property arising under Paragraph 10.6.

4.3.7 Claims for Additional Time.

4.3.7.1 If the contractor wishes to make Claim for an increase in the Contract Time, written notice as provided herein shall be given. The Contractor's Claim shall include an estimate of cost and of probable effect of delay on progress of the Work. In the case of a continuing delay only one Claim is necessary.

13A.2 No action or failure to act by the Owner, Architect or Contractor shall constitute a waiver of a right or duty afforded them under the contract, nor shall such action or failure to act constitute approval of or acquiescence in a breach thereunder, except as may be specifically agreed in writing.

Courtesy of Joseph C. Kovars, Ober, Kaler, Grimes & Shriver

APPENDIX H

Monthly Progress Status Report

MONTHLY PROGRESS STATUS REPORT
AS OF May 31, 2006
Per 01310-3.03 B

1- Revisions to the Schedule: Added 60 calendar days to the schedule per change order 11.

2- Work Completed during reporting period: See attached exhibit A. Items with this Inv % greater than 0 were active this past period.

3- Work to continue or start in upcoming period: See attached exhibit B, which shows items with early start dates within range of data date to 30 days after data date.

4- Problem areas and impact: None more than normal. The check valve issue at the IPS cost us valuable time in that Area. Also, we have lost valuable time at the Chem Feed Bldg, which is now the critical path.

5- Corrective Action Recommended: MAB is working extended hours (until 4-5pm daily) Electric has started working some overtime in May due to delivery of busduct.

6- Effect of changes: None.

7- Updated Tabulation of Contract Time: As of May 31, total float is -165, showing a loss of 22 workdays in May.

8- Evaluation of overall status of the schedule for the job: The Infl Pump Station is now at -146. Pump #4, the second new pump to be installed, was ready for operation April 19th, and testing could have been completed April 24th. Instead, we lost two weeks trying to modify the check valve for this pump and another three weeks trouble-shooting control problems. The valve delay was due to hydraulic conditions and test parameters beyond our scope and control; this was resolved by adding springs to assist the air chamber in closing the valve. The pump control problems were resolved by adjusting the time delay on alarms, as programmed into the PLC controller. No problems were found in either the valve or the pump. We have started work on IPS #3 (existing pump # 2). Hopefully we don't lose further time on similar issues.

The CCT, Chem Feed Bldg and PAB lost 22 days this month. Time lost in the last month in the Chem Feed System is due to delays in completing wiring, calibrating control devices, point-point testing, and network configuration. In addition, time is still being lost due to wiring "clarifications" for the CRP controller (see RFI #1150, 5123106). The startup and testing phase of this system may become more complicated than anticipated, due to the Plant's sensitivity for phasing in the new system while maintaining the existing gas feed.

Pump and Generator Bldg: At the end of May, float is now—126, for a loss of 7 days in the month. Reasonable progress was made this month. Hopeful delivers on their promises for MCC-G delivery.

Courtesy of Joseph C. Kovars, Ober, Kaler, Grimes & Shriver

APPENDIX I

Letter of Delay

December 10, 2004

VIA FAX

Subject: Delay

Dear Bob:

Time is of the essence. You are hereby notified that the delay in utility work has and is causing delay in the progress scheduled subsequent completion of the project. Lack of this activity has already triggered additional work for us and your earth work crew who are forced to perform constant stabilitation and clean-up of the site such as importing limestone dust and mucking out left loose dirt. Lack of manpower not only has delayed other scheduled activities such as access to install concrete on footings, slab, curb, and gutter, but also has limited and restricted storage space for other trades. We request a written recovery plan to include manpower and equipment sent to the job site.

Project Manager

CC:

Courtesy of Joseph C. Kovars, Ober, Kaler, Grimes & Shriver

APPENDIX J

The General Conditions of the Contract for Construction, American Institute of Architects Document

The General Conditions of the Contract for Construction, American Institute of Architects Document A201 (1997), contain the following provision:

4.3.10 Claims for Consequential Damages. The Contractor and Owner waive Claims against each other for consequential damages arising out of or relating to this Contract. This mutual waiver includes:

1. damages incurred by the Owner for rental expenses, for losses of use, income, profit, financing, business and reputation, and for loss of management or employee productivity or of the services of such persons; and

2. damages incurred by the Contractor for principal office expenses including the compensation of personnel stationed there, for losses of financing, business and reputation, and for loss of profit except anticipated profit arising directly from the Work

This mutual waiver is applicable, without limitation, to all consequential damages due to either party's termination in accordance with Article 14. Nothing contained in this Subparagraph 4.3.10 shall be deemed to preclude an award of liquidated direct damages, when applicable, in accordance with the requirements of the Contract Documents.

Courtesy of Joseph C. Kovars, Ober, Kaler, Grimes & Shriver

APPENDIX K

"No Damages for Delay" Clauses

The contractor agrees that the contractor's sole remedy for…delay shall be an extension of contract time and that the contractor shall make no demand for damages or extended overhead. The contractor shall not be entitled to payment or compensation of any kind from the Owner for direct, indirect, or impact damages arising because of any hindrance of delay from any cause whatsoever.

Courtesy of Joseph C. Kovars, Ober, Kaler, Grimes & Shriver

APPENDIX L

CONSTRUCTION PROJECT DOCUMENTATION CHECKLIST

1. All agreements, contracts and subcontracts, including all exhibits and addenda and proposals.

2. General Conditions and any Supplemental General Conditions.

3. Correspondence.

4. Pre-job meeting minutes, minutes of job meetings.

5. Bulletins.

6. Logs.

7. Diaries, phone records.

8. Notes.

9. Memorandums

10. Computer records, including Computer Aided Design drawings.

11. Specifications and/or project manuals, including any addenda.

12. Shop drawing and sample submittals.

13. Approved shop drawings.

14. Rejected shop drawings.

15. Catalog cuts.

16. Approved catalog cuts.

17. Rejected catalog cuts.

18. Plans and/or blueprints.

19. Diagrams.

20. Minutes of safety meetings.

21. Applications for payment.

22. Approved applications for payment.

23. Photographs.

24. Invoices.

25. Work schedules.

26. Progress schedule, bar charts, time charts.

27. Employee time records, payroll records.

28. Progress reports.

29. Survey reports.

30. Bid documents, including requests for information, qualification statements, work sheets, cost projections, labor projections, bid analyses; bid tabulations and comparisons.

31. Project financial data: audit reports, job charge records, project cost accounting records, financial statements.

32. Equipment utilization records.

33. Schedules of values.

34. Work orders, change orders, field orders.

35. "As-built" drawings.

36. Certifications of substantial and/or final completion.

37. Laboratory and/or test reports.

38. Procurement, fabrication, delivery records or reports.

39. Quality control records.

Courtesy of Jonathan D. Herbst, Esq., Margolis Edelstein

APPENDIX M

CONSTRUCTION LITIGATION CHECKLIST

Case Name:
Client Name:
Venue:
Judge:
Mediator:
Trial Date:

THEME OF DEFENSE:

A. Factual Summary:

    1. Name and location of project

    2. Date of completion of project

    3. Description of claims

    4. Client's scope of work

    5. Allegations against client

B. Initial Investigation:

    1. Meet with client: establish goals/objectives of client; establish fees and have frank discussion about fees and costs of construction litigation.

    2. Determine client's scope of work; obtain copies of contract documents; is any portion of the contract oral? ; determine extent of job documentation; obtain a copy of the entire job file; if voluminous, have the documents scanned for computer storage.

3. Obtain chronological sequence of project with emphasis on client's involvement and client's knowledge of claims; obtain identities of key project personnel; obtain identities of other potential witnesses; determine whether a site inspection is possible.

4. Determine whether other parties should be brought into the litigation.

5. Determine whether client has any insurance that may cover all or part of the claims; tender defense to every possible insurance carrier.

6. Contact opposing counsel to obtain case information. What is the theory of liability and what facts (documents / witnesses) support that theory? What is the claim for damages and what facts (documents / witnesses) support that claim?

7. Make follow up contact with client to address unanswered questions or incomplete responses; determine new or unique issues that have arisen as a result of the initial investigation.

8. Determine what experts will be required to assist in the defense effort.

C. Theme:

1. Establish the theme of the case.

D. Conduct Legal Research:

1. Is the theme of defense legally recognizable?

2. What facts must be established to support the defense theme?

3. Is an expert needed to support the defense theme?

E. Follow-Up Investigation:

1. Contact identified witnesses to obtain specific information.

2. Have additional witnesses been identified as a result of these contacts?

3. Review pertinent documentation.

4. What documentation must be obtained from other parties?

5. Have all liability facts that aid the defense of the case been explored?

6. Is the liability analysis supported by the facts revealed in the investigation?

7. Establish a damages analysis.

8. Have cost of repair estimates been obtained?

9. Are there additional damage claim possibilities?

10. Have expert opinions been obtained and reviewed?

F. Revisit Theme:

1. Have facts developed that require a change in defense position?

2. Is the theme viable?

Courtesy of Jonathan D. Herbst, Esq., Margolis Edelstein

APPENDIX N

DELAY CLAIMS CHECKLIST

A. Applicable Contract Language:

    1. Contract scheduling provisions?

        a. Submittal procedure in specifications?

        b. Specification requirements relating to time?

    2. Time deadline for submission of claims?

    3. Time is of the essence clause?

    4. No damages for delay clause?

    5. Coordination of work clause?

    6. Suspension of work clause?

    7. Differing site conditions clause?

    8. Liquidated damages clause?

B. Is the Delay Excusable? Why?

    1. Owner caused delay

    2. Errors or omissions in plans and specifications

    3. Change in design.

    4. Delayed shop drawing submittals caused by others

    5. Labor strike

6. Government action

7. Unexpected site conditions: Is there a contract provision?

8. Weather: Is there a contract provision?

C. Is the Delay Non-Excusable?

1. Not enough work force?

2. Defective work?

3. Delayed shop drawing submittals not caused by others?

4. Financial difficulties?

5. Equipment or material supplier problems?

6. Unexpected site conditions: Is there a contract provision?

7. Weather: Is there a contract provision?

D. Applicable Law:

1. Liability

a. Contract provisions

b. Case law

c. Statutory law

2. Damages

a. Contract provisions

b. Case law

c. Statutory law

d. Concurrent causes

E. Damages:

    1. Field office

        a. Labor costs

        b. Bonding costs

        c. Material costs

        d. Equipment costs

    2. Home office

    3. Owner

        a. Liquidated or actual damages?

        b. Lost profit

        c. Loss of Use

        d. Attorneys' Fees

        e. Interest

        f. Financing

        g. Maintenance and overhead
        h. Goodwill

F. Expert Witnesses:

    1. Construction scheduling

    2. Damages

G. What Scheduling Procedure Was Used?

    1. What was pre-start schedule?

    2. Schedule updates?

    3. Are any schedule updates disputed?

    4. Job progress documentation?

H. Actual Construction Schedule?

    1. Start and finish dates of actual activities

    2. Concurrent causes: What delays extended completion?

    3. Acceleration?

Courtesy of Jonathan D. Herbst, Margolis Edelstein